*Overstory: Zero*, the critically-acclaimed award-winning essay collection by Robert Leo Heilman is back in print in a new revised and expanded edition.

"In prose reminiscent of Wendell Berry, Heilman eloquently captures the struggle to carve out a good honest life."
— *Seattle Weekly*

"He writes of the pain and fear of layoffs, of suicide among friends and neighbors, of a 75 percent divorce rate, of the brutal labor of tree-planting. He also writes of the beauty and solace of the Umpqua River, of . . . little-league baseball, of small-town parades and small-town identity. Heilman records these things simply, with compelling balance, eloquence and compassion."

—Jeff Thompson, *Whole Earth Review*

"A mix of keen observation, social activism and humor, the essays tell a vivid story, unified by Heilman's love for his community . . . Heilman is never satisfied with simple answers, probably his greatest asset as a writer. The reckless ignorance of the weekend warriors who vandalize logging equipment and yuppies who gather at a benefit dinner for hunger draw the same outrage and bemused wonder that Heilman pours out on those who destroy the environment."

—Faris Casell, *The Register-Guard*

"This book won't likely change people's minds about where they stand on managing forests or preserving old growth. It will, however, serve as a lesson in empathy for a hard-working people who have seen a way of life slide away and don't know where or how they will go to work now."

—Susan English, *The Spokesman-Review*

"Robert Heilman is a rugged individualist in a rugged land. He lives what he writes, knows what he thinks, and his works are sculpted by a master's hands."

—Lawson Inada, author of
*Legends from Camp*

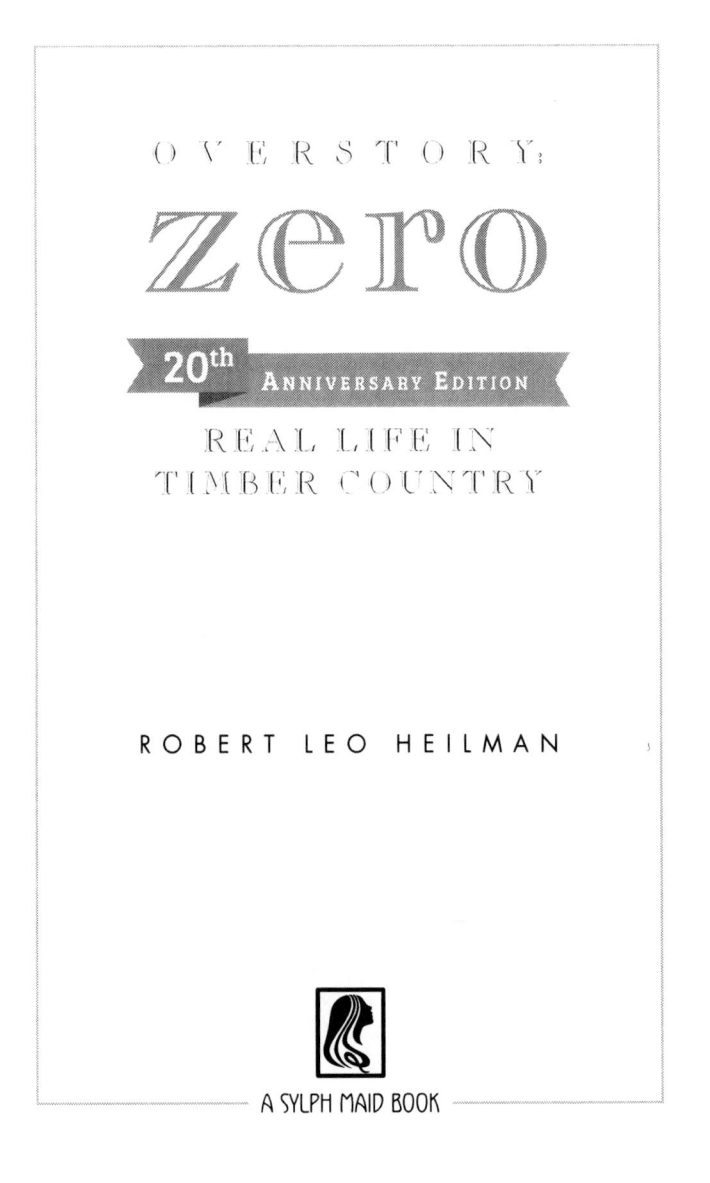

# OVERSTORY: zero

**20th ANNIVERSARY EDITION**

## REAL LIFE IN TIMBER COUNTRY

### ROBERT LEO HEILMAN

A SYLPH MAID BOOK

Sylph Maid Books
P.O. Box 932
Myrtle Creek, Oregon USA
97457
(541) 863-5069

1st edition 1995
2nd edition 2014

ISBN: 978-0-9908686-0-6 (Paperback)
ISBN: 978-0-9908686-1-3 (ebook:Kindle edition)
ISBN: 978-0-9908686-2-0 (eBook/ePub edition)

Cover and interior design, cover and interior illustrations
and composition by Michael Rohani, Design for Books,
www.designforbooks.com

Printed in the United States of America.

"Remember, all of man's happiness is in the little valleys. Tiny little ones. Small enough to call from one side to the other."

—JEAN GIONO, BLUE BOY

For Mrs. Stubkax,
Without whom nothing.

# AKNOWLEDGMENTS

The following pieces first appeared in these publications: "An Occupational Disease" and "Diving High" in *The Oregonian's Northwest Magazine*; "Passes," "Eldorado," "The Smell of Home," "High Water," "Honkers," "Central Heat," "Snow," "Falling Ashes," and "Black Wings" in the *Umpqua Weekly Examiner*; "Of Time and the Gyppo" (as "Death of a Gyppo"), "Turnover," "Of Terror," "High Risk Ground," "At the Zoo," "With a Human Face," and "Why We Celebrate" in *Oregon Quarterly*; "The Milkshed" and "The Politics and Prejudice of Old Growth" in *Northern Lights*; "Overstory: Zero" and "Euclid's Hell" in *Left Bank*; "Getting By," "Degrees of Understanding," "Home of the Vigilantes for Justice," "Rototilling Resurrection," and "Old Timers" in *Jefferson Monthly*; "Small Towns and Quiet Voices" in Writers NW; "The Enemy Among Us or the Enemy Within Us?" (as "Pain of Hard Timber Choices Leads to Fear") in *The News-Review*; "Who Owns the River?" in *Umpqua Watersheds Newsletter*; "The Field of Reality" in *Elysian Fields Quarterly*; "Counting Heads," (as "Counting Oregonians") in *Portland*.

The excerpt on page v from *Blue Boy*, by Jean Giono, is reprinted with permission from North Point Press, a division of Farrar, Straus & Giroux, Inc.

# CONTENTS

I owe my readers an explanation for this revised and expanded edition of my little Umpquan book. After all, it's been nearly twenty years now since Sasquatch Books of Seattle offered me a publishing contract for this collection and a good deal of river water has passed beneath the Myrtle Creek Bridge since then. Myrtle Creek's streets are no longer patrolled by the Vigilantes for Justice, the federal government has abandoned its plan to turn the entire Umpqua National Forest into a maximum-yield tree farm and our county's population has grayed considerably as young families leave and retirees from out of state move in.

Over the years I've continued to write about my home valley here in Southern Oregon. Actually, I've written a number of things over those years besides the newer pieces that I've included in this anniversary edition, but these ten additional essays and memoirs will, I hope, help to update the recent history of this place and add to the portrayal of life here in my home valleys in the first edition.

I remember working on the manuscript for *Overstory: Zero* back in the spring of 1995 and feeling that it was incomplete and wishing that I could just keep adding to it for the rest of my life like Walt Whitman did with *Leaves of Grass*. This place, my home valleys, is interesting enough and complex enough to sustain a lifetime of pondering and description. "The One Hundred Valleys of the Umpqua" it is called and I would like to write one hundred pieces about our lives here, but, having aged another score of years and finding myself looking ahead to "life: the last decades" I doubt now that I'll ever reach that playful goal.

Robert Leo Heilman
Myrtle Creek, Oregon
July 2014

I got stopped by a Douglas County sheriff's deputy last night for operating a motor vehicle that lacked a license plate light. I figured that he must be new around here. After all, I'd once driven my truck for two years with only a single headlight before I was finally told to fix it. He was young, maybe a rookie, probably bored.

After he handed my papers back to me with a verbal warning, I mentioned, out of my bemused puzzlement, that I'd been driving that old pick-'em-up for twelve years and it had never had a license plate light in all that time.

"Well," he admitted sheepishly, "the reason we pull you over for that is to see if you're intoxicated, or driving without insurance or on a wanted list somewhere."

"Gosh," I blurted out wistfully, "it's been a long time since I was any one of those things."

It had been a decade, or more, since a local cop took even a passing interest in me and I was flattered to find myself able to arouse suspicion on the night before my forty-third birthday. Ah, to be young and a suspicious-looking character again . . .

⋙

It's not easy to be seen as much of a threat after you've lived in the same small Oregon town for twenty years, even if you do look a little weird and drive a disreputable wreck. Maybe this book will change that.

I suspect it probably won't and certainly hope it doesn't. But there's always that risk when you present an intimate portrait of yourself and the people and the place you love to a national audience.

"What will the neighbors think?"

That's a question that in many ways defines what it is to live in a healthy, functioning community. Judging by the news from elsewhere, it's a question that fewer people than ever are asking themselves nowadays. It implies not only an obvious fear, but also the profound caring without which no community can long endure. It implies an adherence to local proprieties and a commitment to unmandated but essential duties. No one can force you to be a good neighbor, but there's also no way to be considered one without earning the distinction—no respect without being respectful and therefore respectable.

Respect is a difficult thing to learn and a harder thing to teach. As a baseball coach I found it to be the most worrisome part of the job—infinitely harder than losing two games out of every three, season after season. It's hard for kids to understand the connection between respect and winning ball games. As a man who has also umpired behind the plate, I can assure you that many adults don't seem to understand it either.

Nevertheless, I've had to "call 'em as I see 'em" too many times to do otherwise, whether on a ball field or in an essay. By and large, people respect that. They might not like it, but they respect it—and, if people respect you long enough, they'll eventually forgive you for being wrong or (worse yet) right. They might even like you.

Tolerance is another aspect of real communities, ones that work because the people living in them haven't much choice but to make them work. There is hope too. When you live in a place where you rely on your neighbors and where you can readily see the good or ill effects of your own actions, it's easier to take responsibility for making your home town a better place.

I feel very fortunate to live in such a community. It is a rare thing and I've had a rare chance to examine it and to write about it. My hope is that some good will come from doing that, both for the small world which I and my neighbors inhabit and the larger world of which it is a small part.

I hope that by taking an honest look at our strengths and our weaknesses we can come to care more about our world and learn to respect it and ourselves and each other more. Then maybe—just maybe—if we care enough about each other and this place we call home, we can come up with some gifts that will have a lasting effect on our larger world.

⁂

Just as the world is made of many places and humanity is made of many individuals, so to, every place and every person has a multitude of aspects. All these aspects exist, not in isolation but as pieces of an infinitely complicated whole.

Through these essays, memoirs and sketches, I've tried to portray some of that complexity, to look at connections, between my life and the lives of my neighbors, between our work and our place, our families and our community. It's not a bad place and not a bad life; both are flawed in many ways but blessed in just as many. It's impossible to separate my life from this place, or to separately analyze the uncountable aspects of either.

The aim of all autobiographical writing is apologetics, the saying, too late, of the things that should have been said. It is, of course, a foolish task, born of a desperate desire to be understood. Ultimately, it is a futile one as well, doomed from the outset by an innate dishonesty. Sometimes I think that the only honest writer was Lao Tzu, who had the good sense to admit defeat in the first line of the Tao Te Ching: "Existence is beyond the power of words to describe."

And yet, there is the hope that we may, at times, go beyond mere expression and actually say something useful.

Robert Leo Heilman
Myrtle Creek, Oregon
February 1995

AN OCCUPATIONAL DISEASE

It was the damp, chill autumn time, barely too warm for frost and too wet for comfort. We were working on a Bureau of Land Management stream-cleaning contract, clearing out a log-jam in the Siskiyou Mountains near the California line. My partner, Brian, and I sat up on the stream bank among sword fern and viny maple and waited to see what kind of fool the log would make of the government inspector.

The odds were about fifty/fifty that he'd shortly be a dead fool or a maimed one, and better than even that he'd end up a cold wet one. Regardless of the outcome, we sat in the fog-wet brush near the yarder's tailblock, me smoking a hand-rolled cigarette, Brian with a jab of chew in his cheek, not talking, keeping our thoughts under our hardhats.

Below us and about one hundred feet upstream, the inspector stood where I had stood an hour before, on a wet boulder, looking up at an old rotting log which hung overhead, wedged between moss-dappled rock walls above a small pool. Behind the log a waterfall fed the pool.

Standing there in mid-stream on the slick rock, with the sound of splashing water and the mass of the log above and before me, I'd seen the possibilities and didn't like any of them. If my chainsaw didn't get stuck, if the log's compression didn't send it buckling my way, if I could shift my balance away from the log so that I didn't fall into the pool, if I didn't slip and fall breaking an ankle or rib while scrambling out of the way, I would merely get drenched on a cold fall mountain morning.

From above, up on the bank, it had looked routine; but standing there on that rock I could see that it was lethal. I gave it up as too risky and then Brian walked down there, saw in hand, and came to the same conclusion. "It's funny," he said after climbing back up the bank, "it looks easy from here."

Now it was the government man's turn down there in the hole. He had showed up an hour later, looked down from the bank at the rock and log and pool and declared the log removable. The contract specified a clear, debris-free channel and he was there to make sure we fulfilled the contract.

We refused. "It's not safe."

"Hell, I could cut it out of there myself."

"Okay—go for it," Brian said, and handed him a chainsaw.

There was no use arguing with him, the log had fooled us too, until we stood in the only spot where you could lay a saw on it. If he wanted to prove us wrong, we'd give him the chance. The man might die, as easy as not. The log might crush him, pin him or drown him. We would, of course, try our best to save his life afterwards. But it was his choice now. Anything might happen—and to us it was all the same. Our hearts stayed as gray and featureless as a fog bank.

Though loggers are often portrayed as hard characters, neither of us was cruel or deliberately heartless. Our indifference to his fate could easily be ascribed to machismo, a matter of manful pride, or to class differences with the inspector, whose boast had challenged both our craftsmanship and our courage.

Logging is rough work. Hard labor, long hours, dangerous conditions and male-only companionship almost guarantee a hardening of the heart. There's also the wear and the tear on your spirit from tearing up the world. Work gloves can protect soft hands but tender psyches just develop calluses. Pride and the nature of logging go a long way toward explaining our attitude, but not far enough.

We are all loggers in our own way, though for most of us the brutality and violence of our jobs is more subtle. "'I'm sorry," we say, "it's company policy," as if the rules of corporations were as real and immutable as the laws of nature. Alienation is an occupational disease, one that afflicts each of us when we sell our time for money. It brings a numbness of spirit that makes all sorts of horrible situations seem routine.

At work we become ashen-faced zombies, obediently carrying out tasks whose meaning and effects we seldom care about. We save real living for the weekends. Perhaps there is something in the nature of money itself that poisons all human relations it enters. Or maybe it's something in human nature that leads us to sell off our lives, to trade the possibility of love for a strictly limited security. Whatever the cause, ultimately it whittles us down to its own inhuman scale. Most people are likeable enough away from the job and even at work. We each contain a complexity and beauty beyond the ability of art to portray. We also contain a bleakness of spirit unimaginable. It is in the humdrum, the daily grind, the unreal world of work that we cross between the two without noticing the change.

One hundred feet away, down in the creek bed, the government man stood where Brian and I had each stood in turn. If he tried to cut that log, then he was a fool to doubt us and so whatever happened to him was simply his own doing. We waited and watched as he started the chainsaw and held it at arm's length overhead to start his cut. Wood chips cascaded down into the pool, exhaust smoke mingled with the morning mists. Then he stopped, withdrew the saw, shut it off and came trudging back downstream and up the bank to where we sat.

"'You're right," was all he said, and we were, of course, pleased to hear him admit it.

1987

# PASSES

In the eight hundred or so miles between San Diego, California and Redding, Interstate 5 climbs over one mountain pass, the Grapevine, between Los Angeles and Bakersfield. In the forty miles between Grants Pass, Oregon and Canyonville it takes a roller coaster ride over four passes, Sexton Mountain, Stageroad Pass, Smith Hill and Canyon Creek. Long-haul truckers who are new to the route, mutter over their truck stop coffee and curse the engineers who laid out the road. "When will it end?" they wonder, and "Couldn't they have found a better place to put a freeway?"

Sixty miles farther north and five hills later is the answer to the first question, and no, there's no other way it could have been squeezed through here. Welcome to Douglas County.

Call it ruggedness or just call it sheer orneriness, but that fact is that even the mountains and rivers here do things their own way, running from east to west instead of north and south like the rest of West Coast.

It's a tough place to pass through and an even tougher one to get around in. Neighbors living less than a mile apart but separated from each other by steep ridges, may have to travel forty or fifty miles by road to visit. They might as well be on the backside of the moon.

We feel that we are more than just a little different from other Oregonians. Isolation breeds independence, a habit of mind that brings us some good natured kidding from our neighbors to the north and south about being a backwater place, outside the mainstream of American life, a charge that rolls off the average Umpquan like rain off a mallard.

There's something to be said for living in a backwater. Considering the hodgepodge of fads, fashions, and politics that passes for American mass culture, being different can be a real blessing. We can go on quietly doing things our own way.

The mountains that cut us off also force us together because isolation breeds interdependence as well as independence. Around here being a good neighbor is much more important than your politics or your lifestyle or your religion.

Many folks who live here feel that we should seal off our mountain passes and secede from the state of Oregon and even the Union. But that's not really necessary. The mountains have already given us what we need: the twin gifts of orneriness and neighborliness. The secession has already taken place—in our hearts.

1991

MONDAY MORNING

There's a yellow legal pad sheet of paper hanging on the office door next to the time clock. It lists, in red ink, the work schedule for the week. I find my name under the heading: "Indefinite Lay-off—Subject to Immediate Recall." "Bob Hielman" it says. The asshole didn't even spell my name right. What are the odds that he thought about what that listing means for me and my wife and our baby? Winter, the recession, food stamps and government surplus cheese, Thanksgiving, Christmas, silent worrying nights of waiting for spring and an end to the housing slump. I thought I had it made.

The door that the work schedule hangs on, the wall that frames the door, the shelf that holds the time clock, the lockers behind my back, are all the work of my hands. They are not art but they are well-crafted, functional pieces of industrial simplicity. On Friday I looked at them with satisfaction. Now I just want to escape. I turn and walk out the door without saying a word.

The gas gauge on my 1962 Chevy Nova reads "E" and the starter solenoid won't catch in this cold weather. I sit, clicking the key over and over again, listening to the starter whirr without catching, and wonder if I'll end up walking the mile and a half home.

There's a tapping on my window. It's Tricia and I roll down the window.

"What's up?" I ask.

"Hey, I've got an empty for you in the car," she says, "Wait a minute—I'll go get it."

I've got a one-cow dairy and she's one of my customers. I wait, watching the smoke from the wood burner billowing out and up the side of the building. The stack stops about ten feet short of the roof line and the smoke blackens the light green metal siding. It's a new building and ten feet of pipe could keep it looking new but the stain gets worse day by day, the paint blisters and peels; by spring it'll rust and the triangle of black soot will have a complementary red stain trailing downward. By spring I might have lost my place and the cow with it.

Tricia comes back with a one gallon glass jug. I take it from her and ask if she'll need another gallon on Wednesday.

"Yeah," she says. She's chewing gum and it makes her look like a tough broad. 'Did ya see the notice?"

"Yeah."

"Bummer, huh?"

"Yeah, well . . . that's life, I guess." She's still working. Half the place is laid off and half is still employed. "It ain't the first time."

"No, I guess not." she says, "Jay and Fred are laid-off too." And they too have kids to feed.

"I saw."

"Well, take care." she says, looking a little sad.

"Sure . . . see ya."

This time the starter catches on the first try. The asthmatic engine coughs up a little blood and I'm on my way.

This isn't the first time at all. It is merely another step in an unvarying pattern: hunt for work; find an employer who promises a decent living; work your ass off; get laid-off and start all over again. All for sub-standard wages. Sometimes you get ripped-off worse than other times and once in a while you meet a decent boss you really like. But, in southern Douglas County, Oregon getting laid-off is a fact of life.

I turn left at McCormick Piling's pole yard and start scanning the yard to see who's working and how much wood is on the skids. I see Calvin Poncho cutting knots off of a thirty-five foot log with a double-bit axe, Carl Linde bent over a forty-footer chipping away with his winter spud and Leonard toting three peeled poles behind the jilly truck, backing up to the thirty-footer pile. Leonard is seventy-three and the other two are only ten years younger. Peeling the bark off of telephone pole logs is such brutal work that only old men can do it for long.

Young men are too lazy; the old men work because they must. They have worked hard all their lives and if they ever stop they'll die. They chip away, steadily, patiently, every day, log after bark-clad log, leaving them clean and white for eight cents per linear foot. Calvin could be using a chain saw to cut the knots but none of them would ever do that. They prefer instead the bite of their keen axes.

There are quite a few loads of barkies in right now. If I go in later I might talk them into letting me have a load. They all know me from a summer that I spent working there as a yard man and I'm on fairly good terms with Norman, the farmer who manages the place. My upper back has been bad since the accident last fall and the doctor says I'm supposed to avoid pole-peeling, tree planting, roofing and other stoop labor, but then, I'm not supposed to be doing carpentry either. If I take it easy, I can probably get away with it. We've got to eat, after all.

Across from the pole yard there's a big two-story house with twenty acres of prime bottom land. The house is well-kept and recently remodeled; the land is uncultivated briars and thistle. The place belongs to a likeable middle-aged guy and his wife. He's a top-notch business man but a piss-poor farmer.

Out in the front pasture is an Angus steer named Andy and a Shetland pony named Petunia. Andy's looking a little thin. He hasn't seen a bale of hay or a can of grain since I sold him to the business man last August. At least the fall rains have brought up a little grass now, so he doesn't have it as bad as he did when the summer heat had the place all withered and brown.

The business man owes me four hundred dollars for Andy. He was supposed to pay me on October first, but his wife spent three thousand dollars on new furniture, so he's short on cash. I'll have to go see him today, because I don't have enough money to make my land payment this month. Knowing that it's a trivial sum for him makes it harder to ask for it. I don't want to think about how much I need that money.

The car bumps over the railroad tracks and I make the tight right turn that leads home. Weaver Road is a narrow gravel road that climbs steeply at first and then winds along the hillside above the tracks for three miles before dropping back down to cross the railway again and merge into the interstate. It is lined with oak and Big Leaf Maple that over-reach the road in places forming gold and brown tunnels of foliage.

It is a quiet road, used more by deer and squirrels than by cars. Sitting in my living room I can distinguish the arrival of my neighbors and the mail lady from the passing of strangers by the sounds of their engines. At night, high school kids roar past sliding sideways around the graveled corners and the morning sometimes reveals a pair of panties hanging from roadside bushes. If I had a nickel for every midnight coupling, I wouldn't have to worry about the mortgage.

Diane is surprised to see me home so soon.

"No work today?"

I shake my head. 'Laid off," I tell her, "indefinitely."

"Oh shit!"

She has to go to school. I'll stay home and watch Kurt. We talk about the "dime-a-dip" fund raising dinner for Kurt-o's day care center that we're going to tonight and kiss each other goodbye.

Mr. K is hungry. Usually he eats at the preschool but today he's home. I haven't eaten either, so we make breakfast together, he, standing on a chair, bare-bottomed, making the toast, me, at the stove, frying eggs sunny-side up and venison steak cooked medium.

We sit down to eat at the oval oak table. The eggs are fresh from the hen house, the venison is from a tender spike buck who stepped in front of a friend's gun sights last week, the milk is this morning's, raw and whole, sweet with cream. In the country you can be broke and still eat better than a bank president.

1981

OF TIME AND THE GYPPO

$\mathbb{T}$he look and feel of the wood got to me. Funny how that works, how a rough-sawn pecky-cedar 1x12 in your hands can conjure up memories, put you for a moment right back where you were fifteen and more years ago. I paused for a moment, holding the board but awash in the ghost smells of saw-dust, gasoline, motor oil, rusting iron and damp cotton Lone Star work gloves. My sixteen-year-old son and my nephew, eighteen, stood looking at me.

"This here's some historical wood," I told them. The boys looked puzzled. It was just a pile of old wood to them—a mess they'd rather not be cleaning up. The gray boards lay on the ground, my pickup truck backed up to them. Why this pause, this staring at an old board?

"This is Clason lumber," I tried to explain, "You know Jim Clason. Well, me and his grandpa, ol' Stewart, used to cut this stuff." Stewart Clason was the first logger in the South Umpqua valley to hire me when I arrived from Los Angeles without a clue as to what it meant to work in the woods.

We were in Brian's back yard, over in Tortilla Flats, a Farm Home Administration low-income housing tract that didn't exist back when Brian and I first went to work for the Old Man. The fence posts had finally rotted and the fence had blown down, lain there all summer and now it was September, time to haul it off before the rains came and the wood got too wet for kindling and the ground too damp to drive across the lawn.

'Damn, that was a long time ago," I added, hoping they'd maybe understand a little of what it meant to be handling wood that had been logged and hauled and milled by a gyppo operator who'd died in the woods back before these two boys reached puberty. If they understood, it was only a very little and there wasn't much use in trying to explain it.

'Well, what the hell you guys waiting for? You stand around spacing-out, making me wait. Come on, let's get this show rolling,'Dime holding up a dollar boys!" I tossed the board into the back of the truck, "Stack 'em neat now."

I first worked for Stewart Clason in 1976 when he hired me and a friend of mine to rebuild his mill shed. At twenty-four years of age I'd done enough framing to pass as a carpenter but working with freshly sawn rough-cut lumber milled onsite as needed was a novelty. Actually, nearly everything that happened at S.R. Clason Lumber Company was new to me.

Within a few days I found myself up a draw on some farmer's back acreage loading logs with Stewart's pick-up, an improvised log A-frame loading boom, a snatch-block, a hundred feet of steel cable and a set of large tongs. My job was simply to drive the pick-up forward or back on command without killing my boss. The task was simple but I was nervous with worry about what could happen if my foot slipped off the brake pedal.

Stewart would set the tongs at each log's balance point and then hand-signal me to back up raising the log into the air high enough to allow him to back his flatbed truck in under the log. He'd climb out of the cab and up on the load and swing the log into alignment and then jump down and signal for me to pull forward, lowering the log into place.

He seemed remarkably spry to me, though I learned later that he was only in his mid-fifties, his silver hair making him seem older. His speech was laced with old-fashioned phrases such as "talked to him enough to educate a mule," "not bad for a cull amateur," and "sawing away to beat the band" which added to the impression that he came from an era long past. Later I learned that he'd grown up in the Black Hills country of South Dakota where his father and grandfather had spent their winters logging and hand-hewing railroad ties.

Gyppo isn't a word you'll find in most dictionaries, nor one you hear often in cities. Like many colloquialisms, it started out as a bit of derisive slang, meaning a logging or mill boss who was likely to "gyp" his employees and creditors—someone who operates on a shoestring budget, a fly-by-night operator, a gypsy. Over the years though, the connotation has shifted from the instability and unreliability of small-time operators to the virtues a gyppo needs in order to survive—courage, self-reliance, common sense, hard work, the ability to improvise and an almost pathological optimism.

The woods were once filled with these wily independents, each as "various-minded" and "ready at need" as Odysseus in a hard hat. In the years immediately after World War II, gyppos were the rule in Southern Oregon rather than the exception. Demand for wood was high and the equipment wasn't expensive. Anyone with a bulldozer and a chainsaw or a portable mill and a little luck could set up and make a living—maybe even a fortune.

It was, by all accounts, a too-brief golden age, cut short by periodic housing slumps and a technological productivity arms race. The big outfits drove their undercapitalized competitors out of business by harvesting ever larger amounts of timber increasingly faster and cheaper with increasingly sophisticated (and increasingly expensive) equipment.

The title has grown respectable in direct proportion to the increasing rarity of genuine gyppos. I've heard men with scores of employees and millions of dollars invested in yarders, bulldozers and trucks proudly describe themselves as humble gyppos. Most of them have been the sons or grandsons of gyppos, men who've inherited some of the traditional mindset though not the precarious existence. Certainly, none of them would trade their sweetheart corporate deals and well-financed shows for the old man's hardscrabble assets.

Working for Stewart was, I realize now, very like working in a living history museum. Everything he had was obsolete and either built by his own hands out of old scraps and pieces of discarded equipment or modified almost beyond recognition. His bulldozer, a little gasoline-engined 1931 McCormick-Deering, had a hand-crank starter that required two men to turn it over on cold mornings and only one if the weather was warm. A 1940 Dodge deuce-and-a-half ton truck with a homemade A-frame boom served as his yarder, skidder and log loader. A relatively modern 1958 Ford flat-bed truck with bunks cobbled from scrap steel hauled the necessarily short logs to the mill.

The heart of S.R. Clason Lumber Company was his mill, one of those old circular saw and carriage contraptions of the sort that turn-of-the-century melodrama villains were fond of tying hapless maidens to. Driven by a quirky art-deco Case industrial engine and an improbable arrangement of belts, pulleys and chains, it wasted an incredible amount of wood. With its too-wide saw kerf, he lost a one-inch board in sawdust for every three rough-cut boards it produced. The wonder was that we produced any lumber at all.

Breaking down was more than just likely—it was a way of life. An efficiency expert hired to analyze S.R. Clason Lumber Company would have tossed away his time/motion study clipboard by lunch on a typical day. In such a case, the Old Man would have offered him a cup of coffee from his thermos and said with a grin, "You know, I believe you're right, son. It can't be done. Why, if a man had any sense, he'd give it all up and get him a real job. But then, I never had much sense."

Actually, there was a certain method to his madness. In theory, at least, what he lost in efficiency was made up by what's known in business circles as "low-overhead." The unreliable relics on which his livelihood depended required constant tinkering but no bank loan payments. The constant down-time cut into his profits, but then, the need for profit was smaller than usual. You can get by on very little as long as you don't need much and, more importantly, know how to tell the difference between your needs and your wants.

There also something to be said for being responsible for your own livelihood, hardscrabble as that living may be. What he did every day wasn't really an economic enterprise but rather, an art form. Any fool with no talent and a hundred million dollars can be efficient. But doing good work on a shoestring budget requires both of Thomas Edison's basic elements of genius—perspiration and inspiration. For forty years Stewart did the impossible every day. It was complex, challenging and thoroughly satisfying work. It was also utterly human and humane.

People, it has been demonstrated, just aren't efficient. At best, we can maintain about sixty-percent efficiency at work—thirty-six of every sixty minutes per working hour earning our pay. (The other twenty-four minutes we spend maintaining our sanity.) This is why corporations (which are themselves beings as artificial and nonhuman as any robot) are so fond of machinery and so ready to outplace workers, replacing them with laser scanners, computers and hydraulic cylinders. Ever increasing productivity looks great on a spreadsheet or tucked into a quarterly report, though it's hell for people, families, communities and the land.

What people do best is the impossible, creatively balancing a wide range of conflicting concerns, desires and obstacles to achieve a complex set of goals. When a machine encounters a paradox it ceases to function; people just shrug, laugh and carry on. Researchers in artificial intelligence (an obvious oxymoron) strive to teach a machine to play a decent game of chess. Personally, I'd be more impressed if they could teach one to handle something truly difficult like surviving puberty or divorce.

It is a rare thing in the timber industry of the 1990's for a man and an employee or two to fall, log, haul, mill and sell a particular stick of wood. It would be foolhardy to compete with the specialized links that form an efficient chain in the process. Rare now too are the celebrated virtues of the gyppo, though fifty years ago they were unremarkable, as unnoticed as, say, old growth forest, clean water and healthy salmon runs.

Time, we say, is money, meaning that spending the least time earning the most money is a standard to measure our success. Ultimately, by this logic, the highest success is to spend no time at all doing useful work and to receive more than we could possibly spend for doing so. Only on the Big Rock Candy Mountain.

To the productivity expert, pursuing the corporate dream (if "an agreed upon legal fiction" can be said to dream) of one hundred percent efficient workers, there is never enough time. Squeezing one more marketable unit out of a day only leads to efforts to increase that amount, however fractionally. The hallmark of artists is that they always have plenty of time.

Stewart Clason had time, time to fix whatever was broken at the moment, time to sit talking with an old friend who'd dropped by, time to wait for the weather to cooperate, time for a cat nap every day after lunch, and, always, plenty of time to do a job right.

Doing a good job cost him plenty over the years but he stubbornly insisted on it. I'd hate to characterize him as an idealist, but he did everything conscientiously and derived a lot of pride from that. He was, I think, wise enough to understand that pride was about all his show ran on. Taking a few extra minutes to ensure that he did as little damage as possible to the land came as naturally to him as ensuring his own safety. Time, he had to spare, and so losing a few hours in the course of a job was cheaper, in the long run, than losing his self-respect.

What, in the end, made a quaint anachronism out of him was that he kept doing the same supremely human thing, practicing an art, while all around him logging and milling steadily shifted away from being a unique art form to just another increasingly efficient industrial process.

"It's a trade-off," he told me one unexpectedly balmy January day. "Sure, I could make big money. Work as a millwright maybe, or a supervisor for some big outfit somewheres. But then, I wouldn't have time to sit here drinking coffee and listening to them frogs singing about spring. You've either got the time or the money, but you never have both."

His time is gone now. I guess that's what made me pause for a moment to consider the history of an old rough-cut cedar board that was now useless except to light a hearth fire in my home.

In his time Stewart survived the Great Depression, combat in Europe, a half-dozen severe housing slumps, a hard-fought, violent (and futile) timber faller's strike and the daily risk of death or crippling injury. Using his wits, his muscles and his integrity he supported a wife and four children, taught his two sons the trade he'd learned from his father and grandfather and made himself useful to everyone around him in uncountable, often charitable, ways.

He survived long enough to play with his grandchildren and to become an anachronism. He lived to see his years of struggling against corporate perfidy ignored by a society in which the proud title of logger had become a term of disdain, while those very same corporations used his and his fellow gyppos' lives as an example of what they hoped to preserve by cutting too much timber much too quickly. With old age, he found himself in a world which didn't have a place for him and his kind. And yet, this unkind cut too he accepted philosophically, fully aware of the injustice and ornery as ever. He'd never expected much, never wanted much and never had a whole lot to show for his labors. Nevertheless, he kept working anyway.

We talked of death one day, on a landing up on Tater Hill where we were salvaging cull logs left behind on a clearcut. He'd been down in the hole all morning, bucking logs and setting chokers while I ran the jury-rigged old Dodge that served as our yarder and skidder. A half-hour shy of noon a heavier than usual log hung itself up against a stump, the engine slowed and then stalled. Stewart hand-signaled "slack line" and then "shut-down" drawing a knife-finger across his throat and trudged up to the landing with his chainsaw. "Well, the mule quit, on us" he said, "must be dinner time."

We sat in the shade on the edge of the cut, ate our lunch and talked "of shoes and ships and sealing wax and cabbages and kings" as we did every day. I told a story about a clumsy roofer I'd worked with in New Mexico, who'd hollered "Shit!" as he fell from the edge of a roof, while I shingled unperturbed on the opposite side of the gable thinking he'd hit his thumb again with his hammer.

The Old Man sat stretched out with his back against a tree trunk preparing for a nap. Years of familiarity with sudden violent death, in combat and at work, had reconciled him to the notion that one could be alive one moment and a lifeless mess the next. He confessed though, a deep fear of dying slowly, lingering on in a living death from cancer as many of his relatives and friends had done.

"When I go," he allowed, "I want it to be just like this, stretched out in the woods somewhere on a nice day with my eyes closed."

In the end, six years later, he died of a heart attack while fishing for trout at Crane Prairie Reservoir, alive and joking one moment and stretched out on the dock with his eyes closed the next.

1997

# ELDORADO

His talk is both too proud and too modest and always there's the suggestion that he knows more than he says, though, of course, he'll never come out and admit it. There's never any talk about what he's doing right now, or where, or exactly how or how much, but he's on to something that might or might not be happening or is about to happen, and make no mistake about that. For it's there all right, just like the old timers always said it was—everybody knows that.

It's profoundly rude to inquire too closely or to show any sign of doubt. For all gold miners are are closed-mouthed about their present circumstances. No, it's best to talk about the past and hear the tales of lost lodes and sudden finds, of years of diligent search abandoned just a few short feet away from staggering hidden wealth. It's best just to listen.

Eldorado, the golden one, lives on here surely. You can hear the echo of his footsteps in the mountain canyons—just around the bend. They sound like a shovel digging through gravel, like wet sand sloshing back and forth in a metal pan. You can see his gleaming dust and flakes and nuggets, like pieces of the sun fallen to earth as a gentle rain.

It's nobody's business just how many ounces of the stuff come out of these hills every year—not yours or mine or the tax man's. It comes from no-one-says-where and stays somewhere with someone or goes somewhere else, unmentioned, except as a vague passing hint that there's more, always more, where it came from.

No one can say just how much it's really worth. But whatever it comes to in dollars it's small pickings compared to the fever that grips our imaginings. We can live without the gold but not without the pursuit of it and a mine that stays lost is worth more than one found. All the wealth that might be scraped out from the dark underworld to gleam in the sunlight and delight the eye, is still less precious than the golden dream itself.

1991

## THE MAIN THING

The main thing is to have a big breakfast. It's not any easy thing to do at 4 am, but it is essential because lunch won't come for another seven or eight hours and there are four or more hours of grueling work to do before you can sit down and open up your lunch box.

The kids on the crew, 18 year olds fresh out of high school, sleep in the extra half-hour and don't eat until the morning store stop on the way out to the unit. They wolf down a Perky Pie, a candy bar and a can of soda in the crummy, good for a one hour caffeine and sugar rush. They go through the brush like a gut-shot cat for a while and then drag ass for the rest of the morning.

But if you're a grizzled old timer, in your mid-twenties, you know how to pace yourself for the long haul. You're exhausted, of course, and your calves, hips, arms and lower back are stiff and sore. But you're used to that.

You're always tired and hurting. The only time you feel normal is when you're on the slopes, when the stiffness and fatigue are melted off by the work. It gets worse every morning until by Saturday it takes hours to feel comfortable on a day off. Sunday morning you wake up at four o'clock wide awake and ready to stomp through downtown Tokyo breathing fire and scattering tanks with your tail.

Your stomach is queasy but you force the good food into it anyway, a big stack of pancakes with peanut butter and syrup, four eggs, bacon and a pint of coffee. There is a point when your belly refuses to take any more. Saliva floods your mouth and you force back the retching, put the forkful of food down on the plate and light another cigarette.

It's dark outside and it's raining, of course. They aren't called the Cascades for nothing. It's December and the solstice sun won't rise until eight, three hours and a hundred miles from home, somewhere along a logging road upriver.

Rain coat and rain pants, hard hat, rubber work gloves, cotton liner gloves and a stiff pair of caulk boots stuffed with newspaper crowd around the wood burner. All the gear is streaked with mud except the boots which are caked with an inch-thick mud sole covering the steel nails. The liner gloves hang stiff and brown, the curving fingers frozen, like a dismembered manikin's hand making an elegant but meaningless gesture.

Mornings are slow. It's hard to move quickly when your stomach is bloated, your body is stiff and, despite the coffee, your mind is still fatigue-foggy. You have to move though, or miss your ride and lose your job. You try an experimental belch which doesn't bring up too much half-chewed food with it and relieves the pressure.

The laxative effect of the coffee would send you to the toilet but your ride to town is due soon, so you save it for later. Better to shit on company time anyway, squatting out in the brush. It gives you a pleasant break, a few minutes of hard-to-come-by privacy, and it pisses Jimboy, the foreman, off, since, being a college boy and therefore trained to worry about what people think of him, he could never bring himself to actually complain about it.

Lester the Rat taught him that lesson the first week of the season. Les had just planted a seedling and straightened up and turned his back on the slope to empty his bladder. The foreman glanced back to see him standing there with his back turned and staring idly across to the opposite slope.

"Hey, Gaines, get back to work! Let's go!"

The Rat turned to face him and shook the last golden drops off. He smiled pleasantly, showing a mouth full of crooked snoose-stained teeth. "Sure thing Jim." he said mildly, "You bet." None of the professors up at the university had ever mentioned anything like that and Jimboy blushed delicately while all up and down the line the crew snickered.

Jimboy makes more money than you do and doesn't work as hard, which is bad enough. But he's also afraid. It's his first winter on the slopes and he's not used to riding herd on a gang of brush apes. He also wants to make a good impression on his boss, the head forester, so he tries to push his crew into ever greater production. He sees himself as a leader of men, a rugged scientist overseeing the great work of industrial progress.

Everyone tries to get his goat so that, with any luck, he'll amuse us some day by breaking out in tears like Tommyboy, the last foreman, did. "You guys are just animals!" Tommyboy had sobbed, setting off a delighted chorus of wolf howls and coyote yelps. It was the highpoint of the planting season and a considerable source of pride for the whole crew.

## KAMIKAZES

There's a flash of headlights and the crunch of gravel in the driveway. Mighty Mouth awaits in his battered old Ford. You pull up your suspenders and start slapping your pockets: tobacco pouch and rolling papers, matches, bandanna, wad of toilet paper, pocket watch, jack knife, store stop money—all there. You put on a baseball cap and a plaid woolen over-shirt and gather up your gear: caulks, extra socks, rubber gloves, cotton liners, hard hat, rain gear, coffee thermos and feed bucket and step out into the rain.

There's nothing to talk about on the half-hour drive down the creek and downriver to the mill. You know each other too well by now, riding and working together twelve to fourteen hours a day—two moonlit rides and a picnic lunch every day—for three winters. The Mouth holds a beer bottle between his thighs into which he spits his chew as he drives.

You roll a cigarette and listen to the radio and peer out through the windshield watching for the twin reflection of deer eyes ahead. The road is narrow and winding, the roadside brush thick, and you never know just when a deer will step out or leap, windshield high, in front of you. Every day, somewhere on the drive, you see at least one fresh deer carcass on the road. Headlights dazzle the deer and usually they stand there frozen in their tracks before leaping aside at the last moment. Sometimes they leap toward the headlights though, always a suicidal move for the deer, but, like a kamikaze pilot, they can kill too.

## CRUMMY TIME

The mercury arc lamps light up the mill with a weird, hellish orange glow. Steam rises from the boilers and there's a sour rotting smell everywhere. The huge metal buildings bristle with an improbable-looking tangle of chains and belts and pipes. There's a constant whistling, clanging and screaming of saws and machinery coming from them. Bug-eyed forklifts and log loaders crawl around the half-lit yards, mechanical insects scurrying to keep up.

Through the huge open doorways you can see the mill hands at work in their tee shirts, sorting out an unending river of lumber and veneer into neat stacks. The mill workers sweat like desperate dwarves. They make more money than you and stay dry but you feel pity and contempt for them. The poor bastards stand in one spot all night, moving to the computerized lightning rhythm of conveyors instead of their own human speed. The cavernous interior of the mill sheds seem as cramped as closets compared to the open mountain slopes.

You work for the mill but not in the mill, on a company reforestation crew. Most of the company land is planted by contract crews, but the mill runs a crew that plants land that the contractors won't touch—too steep or too ravaged, too old or too brushy.

Acres away, beyond the log pond, past the five-story tall walls of stacked logs, next to the hangar-sized heavy equipment repair shop, is a small refrigerated trailer full of seedling trees in waxed boxes. Each box contains 600 trees in bundles of fifty.

Mudflap and Sluggo are helping Jimboy load tree boxes into the back of a four wheel drive crew-cab pickup. They are young, straight out of high school, and eager to get a promised job in the mill come spring—if they "work hard and show up every day," of course. So, they help load trees and ride with the foreman every morning.

You transfer your gear over to a mud-covered Chevy Suburban crummy. If you've ever ridden in one you know why they're called crummies. The rig is a mess, both outside and inside. The seats are torn, the headliner is gone, the ceiling often drips from the condensed breath of its packed occupants. But you have a great fondness for the ugly thing. It is an oasis of comfort compared to the slopes.

We spend a large part of our lives roaring up and down river powered by its monster 454 V-8. Of course, none of this travel, or crummy time, as it's called, is paid time. Only the forty hours per week on the slopes earns you money. The other ten to twenty hours of crummy tedium is not the company's concern. Together with the half-hour lunch, also unpaid, we spend eleven to thirteen hours a day together for our eight hours' pay. All winter long we see each other more than we see our wives and children. We know each other intimately after so many cramped hours. We bicker and tease each other half-heartedly, like an old bitter couple, out of habit more than need.

## ARITHMETIC

The ten of us plant about 7,000 seedling trees every day, or about 700 "binos" apiece, enough to cover a little over an acre of logged off mountainside each. It gets depressing when you start adding it up: planting 700 trees per day comes to 3,500 per week or 14,000

per month which amounts to 56,000 trees in a season
for one man planting one tree at a time.

Maybe you've seen the TV commercials put out
by the company: Helicopter panoramas of snow-
capped mountains, silvery lakes and rivers, close ups of
cute critters frolicking, thirty-year-old stands of second
growth all green and even as a manicured lawn and a
square-jawed handsome woodsman tenderly planting a seedling. The commercials make reforestation
seem heart-warming, wholesome and benevolent, like
watching a Disney flick where a scroungy mutt plays
the role of a wild coyote.

Get out a calculator and start figuring it: 700
trees in eight hours means 87.5 trees per hour, or 1.458
trees per minute—a tree punched in every 41 seconds.
How much tenderness can a man give a small green
seedling in 41 seconds?

Planting is done with an improbable looking tool
called a hoedad (or hoedag). Imagine a heavy metal
plate 14 inches long and four inches wide, maybe five
pounds of steel, mounted on a single-bit axe handle.
Two or three sideways hacking strokes scalp a foot
square patch of ground, three or four stabs with the tip
and the blade is buried up to the haft. (Six blows 700
times amounts to 4,200 per day. At five pounds each
comes to 21,000 pounds of lifting per diem and many
planters put in 900–1200 trees per day.)

You pump up and down on the handle, breaking up
the soil, open the hole, dangle the roots down there and

pull the hoedad out. The dirt pulls the roots down to the bottom of the hole, maybe ten or twelve inches deep. You give it a little tug to pull the root collar even with the ground and tamp the soil around it with your foot. Generally, what's left of the topsoil isn't deep enough to sink a 'dag in so you punch through whatever subsoil, rocks or roots lie hidden by the veneer of dirt.

The next tree goes in eight feet away from the last one and eight feet from the next man in line's tree. Two steps and you're there. It's a sort of rigorous dance, all day long—scalp, stab, stuff stomp and split; scalp, stab, stuff, stomp and split—every 41 seconds or less, 700 or more times a day.

The ground itself is never really clear, even on the most carefully charred reforestation unit. Stumps, old logs, boulders and brush have to be gone over or through or around with almost every slash hampered step. Two watertight tree bags, about the size and shape of brown paper grocery bags, hang on your hips rubbing them raw under the weight of the thirty to forty pounds of muddy seedlings stuffed inside them.

Seven hundred trees eight feet apart comes to a line of seedlings 5,600 feet long—a mile and some change. Of course, the ground is never level. You march up and down mountains all day—straight up and straight down, since, although nature never made a straight line, forestry professors and their students are quite fond of them. So, you climb a quarter-mile straight down and then back up, eat lunch and do it again.

It's best not to think about it all. The proper attitude is to consider yourself as eternally damned, with no tomorrow or yesterday—just the unavoidable present to endure. Besides, you tell yourself, it's not so bad once you get used to it.

## Outlaws

Tree planting is done by outcasts and outlaws—winos and wetbacks, hillbillies and hippies for the most part. It is brutal, mind-numbing, underpaid stoop labor. Down there in Hades, Sisyphus thinks about the tree planters and thanks his lucky star every day because he has such a soft gig.

Being at the bottom of the Northwest social order and the top of the local ass-busting order gives you an exaggerated pride in what you do. You invade a small grocery store like a biker gang, taking the uneasy stares of lesser beings as your natural due. It's easy to mistake fear for higher forms of respect and as a planter you might as well. In a once rugged society gone docile, you have inherited a vanishing tradition of ornery individualism. The ghosts of drunken bullwhackers, miners, rowdy cowpunchers and bomb-tossing Wobblies count on you to keep alive the 120-proof spirit of irreverence towards civilization that built the west.

A good foreman, one who rises from the crew by virtue of out-working everybody else, understands this and uses it like a Marine DI to build his crew and drive them to gladly work harder than necessary. A foreman who is uncomfortable with the underlying violence of his crew becomes their target. It is rare for a crew to actually beat up a foreman, but it has happened. There are many ways to get around a weak foreman, most of which involve either goldbricking or baiting. After all, why work hard for someone you don't respect and why bother to conceal contempt?

## BAG-UP

The long, smelly ride ends on a torn up moonscape of gravel where last summer's logging ended. No one stirs. You look out the foggy windows of the crummy through a grey mist of Oregon dew at the unit. You wonder what shape it's in, how steep, how brushy, how rocky, red sticky clay or yellow doughy clay, freshly cut or decades old, a partial replanting or a first attempt. The answers lie hidden behind a curtain of rain and you're not eager to find out.

The foreman steps out and with a few mutterings the crummies empty. Ten men jostle for their equipment in the back of the crummies. The hoedads and tree bags are in a jumbled pile. Most planters aren't particular about which bag they use, provided it doesn't leak muddy water down their legs all day, but each man has a favorite 'dag which is rightfully his. A greenhorn soon learns not to grab the wrong one when its owner comes around cursing and threatening.

It's an odd but understandable relationship between a planter and his main tool. You develop a fondness for it over time. You get used to the feel of it, the weight and balance and grip of it in your hand. Some guys would rather hand over their wives.

The hoedad is a climbing tool, like a mountaineer's ice axe, on the steeper ground. It clears the way through heavy brush like a machete. You can lean on it like a cane to help straighten your sore back and it is the weapon of choice when self-defense (or a threat) is needed. It allows you to open up stumps and logs in search of the dark gold pitch which will start a fire in a cold downpour and to dig a quick fire trail if your break-fire runs off up the hill.

The foreman hands out the big waxed cardboard boxes full of seedling trees. The boxes are ripped open with a hoedad blade and the planters carry double handfuls of trees, wired up in bundles of fifty, over to the handiest puddle to wet down their roots. Dry roots will kill a tree before it can get into the ground, so the idea isn't purely a matter of adding extra weight to make the job harder—though that's the inevitable result.

Three to four hundred trees get stuffed into the double bags, depending on their size and the length of the morning's run. If the nursery hasn't washed the roots properly before bundling and packing, the mud, added water and trees can make for a load that is literally staggering.

No one puts on their bags until the boxes are burnt. It is an essential ritual and depriving a crew of their morning fire is, by ancient custom, held to be justifiable grounds for mutiny by crummy-lawyers everywhere. Some argue that homicide in such a case would be ruled self-defense, but so far no one has ever tested it.

The waxed cardboard burns wonderfully bright and warm. A column of flame fifteen feet high lights up the road and everyone gathers around to take a little warmth and a lot of courage. Steam clouds rise from your raingear as you rotate before the fire like a planet drawing heat from its sun. It feels great and you need it, because once the flames turn to ashes you're going over the road bank.

"OK. Everybody get loaded and space-out," The Mouth calls out. You strap on your bag, tilt your tin hat and grab your 'dag. You shuffle over to the edge of the road and line up eight feet from the man on each side.

## In The Hole

The redoubtable Mighty Mouth, the third fastest planter, plants in the lead spot and the men behind him work in order from the fourth to the eighth fastest men. It is a shameful thing to plant slower than the guy behind you. If he's impatient, or out to score some brownie points with the boss, he'll jump your line and you plant in his position, sinking lower in the Bull-of-the-woods standings. Slow planters get fired and competition is demanded by the foreman.

There are many tricks to appearing to be faster than you really are—stashing trees, widening your spacing, pushing the man behind you into the rougher parts while you widen or narrow your line to stay in the gravy—but all of these will get you in trouble one way or another, if not with the boss then, worse still, with the crew.

The notion is to cover the ground with an eight-foot by eight-foot grid of trees. If mountains were graph paper this would be easy, but instead, each slope has its own peculiar contours and obstacles which throw the line off. Each pass, if it follows a ragged line, will be more irregular than the last pass, harder to find and follow. It is difficult enough to coordinate a crew strung out over a hillside, each planter working at a different rate, going around obstacles such as stumps, boulders, cliffs and heavy brush, without compounding it by leaving a ragged unmarked line behind for the next pass.

The two fastest planters, the tail men, float behind the crew, planting two to ten lines apiece, straightening out the tree line for the next pass. They tie a bit of blue plastic surveyor's tape to brush and sticks to mark the way for the lead man when he brings the crew back up from the bottom.

## CUMULATIVE IMPACT

It's best not to look at the clear-cut itself. You stay busy with whatever is immediately in front of you because, like all industrial processes, there is beauty in the details and ugliness in the larger view. Oil film on a rain puddle has an iridescent sheen that is lovely in a way that the junkyard it's part of is not.

Forests are beautiful on every level, whether seen from a distance or standing beneath the trees or studying a small patch of ground. Clear-cuts contain many wonderful tiny things—jasper, agate, petrified wood, sun-bleached bits of wood, bone and antler, wildflowers. But the sum of these finely wrought details adds up to a grim landscape, charred, eroded, and sterile.

Although tree planting is part of something called reforestation, clear-cutting is never called deforestation—at least not by its practitioners. The semantics of forestry don't allow that. The mountain slope is a "unit," the forest a "timber stand," logging is "harvest" and repeated logging "rotation".

On the work sheets which the company's foresters use is a pair of numbers which track the layers of canopy, the covering of branches and leaves which the living trees have spread out above the soil. The top layer is called the overstory, beneath which is a second layer, the understory. An old growth forest, for example, may have an overstory averaging 180 feet high and an understory at seventy-five feet. Clear-cuts are designated "Overstory: Zero"

In the language (and therefore the thinking) of industrial silviculture, a clear-cut is a forest. The system does not recognize any depletion at all. The company is fond of talking about trees as a renewable resource and the official line is that timber harvest, followed by reforestation results in a net gain. "Old-growth forests are dying, unproductive forests—biological deserts full of diseased and decaying trees. By harvesting and replanting we turn them into vigorous, productive stands. We will never run out of trees," the company forester will tell you. But ask if he's willing to trade company-owned old-growth forest land for a reforestation unit of the same acreage and the answer is always "No, of course not."

You listen and tell yourself that it's the company who treats the land shabbily. You see your frenzied work as a life-giving dance in the ashes of a plundered world. You think of the future and the green legacy you leave behind you. But you know that your work also makes the plunder seem rational and is, at its core, just another part of the destruction.

More than the physical exhaustion, this effort to not see the world around you tires you. It takes a lot of effort not to notice, not to care. You can go crazy from lack of sleep because you must dream in order to sort out everything you see and hear and feel during the day. But you can also get sick from not being truly awake, not seeing, feeling and touching the real world.

When the world around you is painful and ugly, that pain and ugliness seeps into you, no matter how hard you try to keep it out. It builds up like a slowly accumulating poison. Sometimes the poison turns to venom and you strike out, at work or at home, as quick as any rattlesnake, but without the honest rattler's humane fair warning.

So you bitch and bicker with the guys on the crew, argue with the foreman and snap at your wife and kids. You do violent work in a world where the evidence of violence is all around you. You see it in the scorched earth and the muddy streams. You feel it when you step out from the living forest into the barren clear-cut. It rings in your ears with the clink of steel on rock. It jars your arm with every stab of your hoedad.

## THE LONG MARCH

"War is hell," General William Tecumseh Sherman said, because, unlike a Pentagon spokesman, he was in the midst of it and could not conceive of something so abstract as "collateral damage."

"Planting sucks." we say, because unlike the mill owner who signs our paychecks, we slog through the mud and bend our backs on mountain slopes, instead of reading progress reports on reforestation units. Like infantry we know only weariness and hopelessness in the face of insanity.

"The millions of trees that the timber industry plants every year are enough to plant a strip four miles wide from here to New York," the foreman tells us.

Our hearts sink at the thought of that much clear-cutting but Madman Phil, the poet, sees a vision.

"Forward men!" he cries, "Shoulder-to-shoulder we
march on New York! The American Tree Planter! Ever
onward!"

Someone starts it and then the whole crew is hum-
ming "The Battle Hymn of the Republic" while, in our
minds, we cross the Cascades, the Snake River Valley,
the Rockies, the Great Plains and onward, ever onward,
a teaming, faceless coolie army led by Walt Whitman,
Sasquatch and Mao Tse Tung, a barbarian horde leaving a swath of green behind us "from sea to shining sea"

"Oh God!" Jimboy moans, "You guys are crazy."

1993

## EUCLID'S HELL

During the Watergate summer of 1973, while Sam Ervin roasted Nixon administration witnesses, I worked as a roofer on a housing development in New Mexico. The days had an amazing sameness. The one-hundred-degree-plus weather held for weeks on end. Though there were six different floor plans for the housing units that we were building, they had only two roof styles—one with a skylight in front and one without. The shingles were either light or dark brown. Each roof took two days to lay. Every measurement, every vent, and each piece of metal flashing was the same as the roof before and the roof that followed. The gravel-coated asphalt shingles formed a Euclidean hell more arid and featureless than the surrounding desert.

Every day a certain cloud would form over the same peak of the Jemez range in the distance. When it grew to the right size, I would confirm with my pocket-watch what the cloud had already told me: lunch time was at hand.

In the relative coolness beneath the roof we ate our meal with the assembled hard-timers, hippies, Chicanos and Indians who made up the construction crews and listened as the contractors argued. There were frequent arguments, sparked by trade chauvinism, conflicting schedules and methamphetamine.

One day the head electrician and the framing foreman got into it. The electrician drew out the blueprints for the house and pointed from the plans to the wall and back again saying: "See? It calls for a doubled stud right here. How the hell can I hang a box here unless you double it up?"

The carpenter was nonplussed. "This place was screwed-up from the get. The foundation's off; the slab's wrong. Face it: It wasn't built to the plan—it was built to the hill. You've got to make allowances."

Despite the fact that the carpenter really should have doubled-up that stud, I remember feeling that there was something important about the exchange without understanding its significance. Like a Sufi story, the carpenter's complaint came back to me over the years, always ringing true, but only slowly revealing its implications.

It was, I believe, the ancient conflict between what is and what ought to be, between the vision and the reality, between mind and matter, with mind stubbornly insisting that its expectations be met and reality even more obstinately refusing to be something it isn't. "Wish in one hand and shit in the other—see which one fills up first," goes the proverb.

We were, as craftsmen, caught up in a no-win situation. Someone, or some group of somebodies, somewhere, had created a plan, a vision set forth on paper, as clean and abstract as a problem in geometry. A housing tract, consisting of housing units, would rise on some lots. Every detail had been considered beforehand. It simply remained for us to follow the dictates of blue lines on white paper.

And yet, that paper village could never stand on this earth. Nature, both human nature and Mother Nature, ensured that.

Stubborn reality refused to conform to the unreal desires of mind. Each piece of the plan, when put into execution, asserted its own individuality against the mind that treated it as undifferentiated, interchangeable parts. No two houses could really be the same. No two nails, of the kegs we pounded, were identical; no two boards or shingles or grains of sand in the concrete were truly the same as any other that ever was or would be. A common everyday miracle prevented, once again, the drabness of human thought from reproducing itself. Walt Whitman would have been pleased.

It's amazing to me how little respect most people seem to have for reality. The mind is a wonderful thing and perhaps most wonderfully of all, capable of tricking us into accepting its version of what takes place around us. We mistake our perceptions for the stuff of existence, repeatedly, even when we know better. Like a kitten trying to touch its image in a mirror, we reach out to the world we think we see, only to find that it's not really there.

I know many people who are terrified by the notion that reality is, by its nature, incomprehensible. A very few are delighted by it. Most people, it seems, never take up the question at all.

I have heard and read about some of the reasons that so many of us trust our perceptions more than we trust the world as it is. It's difficult to sort them all out and get a clear picture, but it's not hard to see what happens when plans seem more real to us than what is actually there. Every act of mass destruction has a logical, and often noble, goal.

❧

In the spring of 1980 I was finishing up my fifth winter as a tree planter in southern Oregon. It was my sixth crew in five seasons and by that spring I'd planted 150,000 seedling trees, enough to replant something close to three hundred acres of logged off mountain slopes. At ten planters to a crew I must have helped reforest about 3,000 acres, all within an eighty-mile radius of my home.

One morning we were planting some freshly clear-cut land up Buck Creek, a three-hundred-acre rocky, ravaged reforestation unit. Our crew was attempting to plant Douglas fir seedlings with ten-inch roots in a perfect eight-foot by eight-foot grid pattern in shallow, eroded soil and logging debris.

Over and over again, we bent our backs and swung our hoedads only to clink against rock covered with three or four inches of topsoil. We did what we could for the land, finding small pockets of soil built up on the backside of the huge, sap-oozing stumps of the forest that used to be and scratching holes in the shallow spots as proof that we'd been there and found the spot unplantable.

"Shit!" someone spat after yet another arm-jarring clink, "I feel like a goddamn chicken scratchin' around out here."

"Buck-buck-buh-gawk! Buck-buck-buck-buck-buh-gawk!" I answered, and the rest of the crew took up the call.

It was spring, a beautiful, sweet-smelling sunny day and a sort of madness, compounded of ten wiry bodies in motion, sunshine, and the frustration of the work, overcame us. We glanced over our shoulders slyly, challenging Jack, the foreman, to stop our clucking insurrection.

Jack surveyed the scene from his stump-top roost, leaning against his inspector's shovel. He lifted his hard hat, scratched his head and decided to try to change the subject. As was his wont, he spoke of the wonders of modern forestry.

"Boy, they sure did a nice job on this unit. Lots of reprod." He gestured toward some scraggly residual trees. "The loggers sure pissed and moaned when we made 'em get good suspension but those naturals will really take off now that we let the sunlight in. In the old days we wouldn't have bothered, you know. Hell, ten years from now this'll all be thick as dog hair with young firs."

"Fuck-fuck-fuck-fuck-off!" someone down the line called out amid the clucking and chinking. Jack pretended he hadn't heard.

I thought of the hardscrabble canyons of Rock Creek, of the old units logged twenty and thirty years ago that we'd replanted all winter long, trying for the fifth or sixth time to bring back the forest on land whose soil had been muddying the river for decades. Something sad and ugly rose within me. I stood up, leaning on my hoedad to straighten my sore back, hitched up my tree bags to ease the chafing on my hips and turned to face my foreman.

"You're nuts," I told him, "This is totally fucking insane." I gestured downstream at the silted creek bed, at the place where we'd found chips of jasper knapped from tool cores by the Indians, at the stark gray face of Buck Rock gleaming in the sunlight for the first time in ten thousand years, at the yarder tower roaring its diesel roar and hooting like an owl as it dragged a turn of logs uphill to the waiting log trucks.

"Look at this place Jack. This ain't a forest—it's a disaster. Get your head out of your ass and look at it. A hundred years from now people will wonder how the hell we could have been so fucking stupid."

The clucking had stopped. The rest of the crew stood still, grinning and watching and waiting to hear Jack's reply.

"This is good ground. This unit will come back just fine. We've done a good job logging it, the best anybody can do, with all the best techniques we've got and it'll come back just fine."

It was time to back off. I knew that. It was one thing to bait the man as a joke but challenging his profession was stepping over the line. I couldn't back off though. The accumulated poison of five winters of tree planting had turned to venom.

"Bullshit. You're fucking crazy. All you company foresters are insane. Just look at this place Jack. Take a look around you and see what's really going on here. It's totally insane."

"Look, Heilman, don't fuck with me. I run a good crew and we do good work. What do *you* know anyway? Huh? I've got a masters degree in forestry—I know what I'm talking about. You don't know shit," he said, as if mountains were blackboards.

I'd blown my next winter's job, his tone said. I thought of November and the uncertainty of finding another crew to work on. In the ten years since dropping out of high school I'd been laid off, fired from, or quit thirty different jobs.

"Yeah, what do I know? I'm just a dumb-ass tree planter."

"Shut up and get back to work."

I glanced at the last seedling I'd planted, chose a likely spot eight feet away for the next one, took two steps and swung my hoedad. Up and down the line the laughter and clucking had died and the only sounds were the scraping and clinking of hoedads on rocks and the distant roar of the yarder.

Jack wasn't a bad guy to work for at all. In fact, I liked him and respected him a good deal. It's not easy to ride herd on a bunch of mud-spattered brush apes and he did it well. But like a lot of nice people, he'd bought into a plan, some words on paper which he never questioned despite the evidence all around him. In his view, the plan itself was foolproof. If anything went wrong it had to be because the plan hadn't been executed properly. It never occurred to him that no plan, no matter how detailed, could ever encompass something as complex and miraculous as a mountain slope.

The lack of respect for the fact of individuality makes all sorts of horrors and cruelties not only possible but seemingly desirable. After all, if the universe is composed of interchangeable pieces, the annihilation or impoverishment or demeaning of any one piece, whether a rock or a mountain, a tree or a forest, a person or a people, a valley or a planet, cannot have much importance.

The underlying integrity of the cosmos is its infinite individuality. You can call it the thumbprint of the Creator, or simply mutter along with the carpenter that "You've got to make allowances," but the important thing is to honor it.

The illusion of sameness creates a devalued currency in our language, thoughts and emotions. We forget that the word only stands for the thing suggested, and the object itself is, by its essential nature, unknowable mystery and sacred in and of itself—simply by being.

Uniformity is a convenient fiction, useful for fooling ourselves but useless for seeing things as they really are. Never trust anyone who believes in the reality of units. They have sold their share in living for counterfeit coinage.

1992

# THE MILKSHED

*A guti Kuhe deckt alle Armut zue.*
*A good cow hides all poverty*
—ALSATIAN PROVERB

It's dark out, wet and cold December. There are no stars, just a fine mist of rain, Oregon dew, settling on the pasture, dripping from the barn roof. The first signs of the solstice sun will be just beginning to show soon. But not yet, not until after the milking. For now, there is only this shadowless gray world, not quite darkness, not quite light.

If you pause on your way, between the house and the barn, milk buckets in hand, you can be lost for a moment, unsure of time, unsure of where you are walking, here in the silently drifting mist between earth and sky, between waking and dreaming, the past and the future. Here, between the house and the barn, on this, the longest night of the year, there is an overwhelming sense of connection, despite the solitude. Somehow, the night is made more real, despite the vagueness of a world without shadow or horizon—perhaps, because of it.

The woman and the little boy are asleep behind you. The old cow and her small, perfect daughter lie in the dark ahead, their backs to the same wall, sharing their warmth through the thickness of the boards between them. The cock roosts among the hens in the chicken house. The dog lies curled on the hay. The dark and the mist and the silence make them more real and more close and more precious than the light of day. They are revealed by the darkness, at once both the well-known and deeply loved particular beings you share this world with and also, every woman, every child, every cow and calf and cock and hen and dog past and present and yet to come.

And what if they all somehow weren't? What if the old essential rituals have finally been neglected and the sun really has died this time, and the great cycles have ground to a halt, come apart at last, with no connections between them anymore? Broken and shattered into its logical components, the fragments of reality would whirl chaotically, each piece alone and isolated and unfeeling, their truth retained as unrelated strings of random digits. Without compassion the night would go on forever.

But there are mouths to feed and they depend on you. You don't need to look far for a prayer that will hold the world together. This simple chore is an essential ritual.

There are rhythms in the night. Your own breathing, your thumping heart and the pulsing of blood, assure you and reassure you. This daily act is all the sacrifice it takes. This walk is the necessary graceful dance which keeps it all going, ". . . world without end. Amen."

There is a stirring in the barn. You reach for a stick-match, strike it and light the lantern. The dog is already risen from the straw, stretching, and then she smiles up at you. Overhead, the cats, resting from their hayloft mouse hunt, meow. The old cow exhales a sweet grassy breath, rubs her horns on the wall and her wide-eyed calf turns her head, regards you calmly.

Each familiar movement is informed. You hang the buckets up on the twenty-penny nail, leave them dangling from the post. Your rubber boots sound the ladder rungs as you climb to the loft, one hand carrying the lantern, one hand pulling you upward. You pass through a hole in the floor, one knee touching the boards, genuflecting, as you step off the ladder into an open space between piled bales of last summer's grass.

Crossing the floor, the cats, a sleek and stealthy mother and her fat sluggard son, rub against your legs with plaintive cries and purring. Down below, you hear the cow mounting the wooden platform of the stanchion, her hooves sounding. She pisses on a flat rock like a hard rain.

You look to the feed sacks, the stacked bales of alfalfa and grass hay anxiously. How long, you wonder, can this small world without money go on? The sun set yesterday on a breaking valley, shattering from the hammer blows of a collapsed economy. There is no work left but this. Across the valley, in a government housing tract a dozen families—mothers, fathers, children—lie asleep. What if the sun doesn't rise? What if the cow goes hungry, runs dry, and they no longer have a jug of milk, at least, to calm their bellies now, when there is no way to earn their daily bread?

It is important to do this right. Each vital step in the dance against death comes in its proper order. Done right, there is no strain or wasted movement. You drop hay flakes down the chute to the feed-rack below; fill the grain tub, measuring the days from the sagging feed sack with a coffee can scoop; place a flake of hard green alfalfa atop the tub; walk, with both hands full, balanced, down the ladder to the dirt floor below.

She stares at you, head tilted with one dull brown eye beneath a curving horn and moos expectantly as you enter the milk-shed carrying a pool of lantern light. Her gold and white sides, brindled with black stripes over the gold, heave with her anxious breathing. Her pink teats stand out sideways from the bulging white bag.

Mindful of her bulk, you set the grain before her. Mindful of her sharp horns, you duck aside as she thrusts her head through the gap, slide the hinged board in behind her ears and lock it in place with a block of wood.

Hearing the click of wood against wood, the dog squeezes under the gate and, after circling around in the still-warm straw of the cow's bedding, lies down in the hollow to watch. You pull up a stool and swiftly perform an ablution, knock the straw and mud and manure from her bag, wash the teats with warm bleach water and a rag, cover your hands with bag balm ointment, anoint the teats and place a bucket between your knees.

You tap her rear leg and she moves it back for you, swatting you familiarly with a flick of her tail out of mild annoyance. You press the side of your face against her warm solid side, your nostrils filled with the primal scents of cow and manure and fresh hay and you are grateful for this moment.

You grasp the far teats and begin the rhythmic squeezing, your forearms and hands sore and stiff at first. Left. Right. Left, right; left, right. Each squeeze produces a response as the good milk washes against the bucket, high pitched and hollow at first, then deeper tones as it fills with the steaming white fluid.

You get lost in the rhythm after a while. The left hand is peace; the right hand is love. Left, right; peace, love. It becomes part of your breathing then, inhaling peace, exhaling love; inhaling love, exhaling peace as the cow's ribs too expand and contract. Time no longer exists in the immensity of a moment. You might as well be, and perhaps you are, any one of your ancestors stretching back to Neolithic times, feeling the same ancient pulse and rhythms, smelling the same scents, waiting for the same sun. And then, off in the hen house, the cock crows.

1996

GETTING BY

"While the 1979–1985 years were disappointing for wood products manufacturers and workers alike, the net result may well have been a blessing in disguise for the survival and long-term health of the industry. Hundreds of inefficient sawmill, veneer and plywood plants have been shut down. Others have been modernized and streamlined to achieve maximum return per unit of product. New products have been developed which require lower labor costs, cheaper raw materials, and result in a better value to end-user. The net effect is a leaner, more productive and cost-effective industry . . . One of the principal means of reducing costs, of course, is to lower wage rates."

— DEVELOPMENT REPORT AND PLAN
CCD BUSINESS DEVELOPMENT CORP.
ROSEBURG, OREGON, JULY 1986

I used to work for a white-haired, old gyppo logger who taught me many things about living and working around here. I remember one morning when he told me about the recession of 1958. We were taking a coffee break in the little two-man sawmill he'd built with discarded equipment and old truck parts. As he talked, we watched a sparrow building her nest in the rafters of the mill shed.

It was 1978 and timber industry wages and per capita income levels in Douglas County, Oregon, where I live, were at an all-time high.

"They used to call them panics," he said, "and then it was depressions, and nowadays they call them recessions, but it's all the same thing. Every time things get rolling good, to where there's lots of small outfits working, money gets tight and the bottom falls out of everything. Then, after the smoke clears, you look around and most all the little guys are gone and the big outfits are bigger than ever."

That particular slump, back in 1958, had cost him his home and his sawmill. He went to work in Northern California as head sawyer of someone else's mill. He stayed a few years, long enough to gather up a grub stake and return home to begin again. "A man does what he has to do to get by," he said.

~~~

In 1982, during Ronald Reagan's experiment in supply side economics, the "trickle-down" recession, unemployment in Douglas County, Oregon, rose to 17.2 percent and per capita income dropped to 77.4 percent of the national level.

There are no 1982 figures available for the rate or amount of emergency food usage because there was almost no emergency food distributed in Douglas County. There were about 93,000 people living here then, as now, and they probably would have eaten nearly a million pounds of emergency food that year, as they have every year for the past seven years. But there was no emergency food, only the need.

Actually, numbers, even if available, wouldn't tell the truth—not enough of it anyway. Besides, I don't trust numbers. People care about what they can understand in their hearts and only machines, or perhaps machine-souled people, can take numbers into their hearts.

The tragedies of humanity are every day tragedies. They can't be expressed as percentages or as issues. They are real and personal individual sufferings and deserve to be spoken of in a way that numbers and jargon can't be used for, that is to say, one should tell the truth, something numbers and jargon never convey.

In the spring of 1981 I sat sharing a 49-cent quart of beer with a friend of mine. It was about eleven o'clock in the morning in the middle of the week but neither of us had anything better to do. We were both out of work. He'd been laid off and I'd been disabled by a fall from a house roof.

From the front porch of his rented mobile home we could see across the trailer court to the creek where the willows wore their new greenery. His daughter and my son, both preschoolers, played on the lawn. The buzzards had returned to the valley making the arrival of the new season official and we drank and talked about the economic slump and the prospects of finding work for the summer.

"My dad told me the other night that there's a plan for this area," he said as we passed the bottle back and forth. His father was an accountant and a local business consultant. "He says the mills are going to stay shut until everyone who can afford to move leaves. The mills are all automating and whoever's left will be the people who can't go anywhere else and they'll work cheap. It's going to be like Arkansas or Mississippi around here— no more good wages."

"It wouldn't surprise me," I said, just to be agreeable, although I honestly thought I was hearing yet another conspiracy theory. Something in my tone of voice must have betrayed my doubt.

"No, really," he insisted, "I don't mean to say that this recession's all been rigged. They're just going to take advantage of it is all. The owners are all going to use it to get people to work cheap. Once everybody gets hungry enough we'll all take pay cuts just to go back to work. They're talking about 'cost-effectiveness' and 'competitive wages' and like that."

There was an inescapable logic to what he was saying, almost like an algebraic equation: Hungry people work cheap; people who work cheap go hungry.

"Well," I said, "You're probably right. I don't know. Nothing we can do about it anyway. We'll just have to wait and see how it all turns out."

That summer I bought a Brown Swiss milk cow named Marygold and raised a flock of Rhode Island Red laying hens. Every day Mary gave about four gallons of milk and the hens laid about a dozen eggs. I began selling the milk and eggs to my friends at cost, charging enough to cover the feed but not for my labor. I found a baker who wanted raw milk and fresh eggs for his family and began bartering a dozen eggs for a loaf of bread or two loaves for a gallon of milk. I had a dozen families on my route, all friends of mine, all young, all married couples with children.

I let them have the food on credit and allowed them to pay in barter (poached venison a few times) or in food stamps when they didn't have the cash. Somehow there was always enough cash for a bale of alfalfa hay or a sack of grain when I needed it.

All this was illegal, of course, but laws aren't something you consider when you're hungry. Poor people break laws as a matter of survival; corporations break laws as a matter of business acumen. Like most of my neighbors (and all of my friends) I lived as the pettiest sort of criminal—driving without car insurance; selling raw, uninspected milk and not reporting the income; accepting food stamps without authorization; cutting firewood without a permit; eating poached salmon and venison.

It's not something I'm proud of, nor particularly ashamed of either. Pride and shame were luxuries we couldn't afford at the time. Many times the jug of milk and carton of eggs that I dropped off was all a family had to eat. We were friends. We had children. We did what we had to do to get by. Even so, most of us didn't survive with our families intact.

When the unemployment insurance checks ran out our little towns started to empty. "For Rent" signs appeared in house windows on every street. Families sold everything that wouldn't fit in the back of their pickup trucks to raise money for the trip to wherever the rumors said there was work.

Dan and Joy held their moving sale just before Christmas and left for Texas with their three sons right after the New Year. Dan had been down to Houston to scout it out and was happy to be doing something—anything—for a change.

"It's not like around here," he told us, "There's all kinds of work, not just the woods or the mill. The pay's not as good but there's steady jobs and all kinds of stuff for the kids to do, concerts and parks and stuff."

It was their last day in the valley and all of their friends were over. They stood among their packed pos-sessions giving away house plants and knickknacks that hadn't sold. I pulled my pickup alongside the house and we loaded up the remainder of the firewood.

"That goddamn Bob," Dan laughed, "instead of going out and cutting him a load, he just waits and burns your wood after you're gone."

We all laughed because it was true.

"That's OK," he added, "I won't need no fuckin' firewood in Texas."

They left the valley the next morning and by the following Christmas they were divorced, Dan was managing a fast food place in Dallas and Joy was living on the streets of downtown Houston and sleeping in an all night movie theatre.

Despite the larger horrors there was much beauty too in the little things we did, uncountable acts of kindness and love, that made the times more bearable, but which in the end weren't enough to keep us going on.

Bit-by-bit we lost our self-respect as we knocked up against hard realities and even harder institutions and agencies that were indifferent to our humanity. It was a slow process, one that we feared but couldn't really see happening.

It's hard to wait when you're used to working. There's a slow steady erosion that wears down a few people first, and then whole families crumble like dirt clods running through your fingers. Finally the community itself is gone, washed downriver, never to return.

One-by-one the families broke up—nine out of twelve couples divorced within six years. Three of my friends, all men, died violently.

Numbers again, cold and flat, devoid of sorrow or joy, of dignity or beauty. Where is the formula that can calculate the supply and demand of affection? We can easily show that x amount of dollars spent on wages lowers profitability by n%, but where is the algebra to calculate human suffering?

I'd like to think that the survival of myself, my wife and my son, was due to right living and perseverance on our part, that somehow we got by, when others didn't, because of our individual and familial strengths. It would be a comforting notion, if I could only believe it. But I'm convinced that it was only the result of a statistical fluke.

At haying time, in the early summer of 1984, I attended a presentation put on by The Hunger Project. The physical therapist who was helping me out talked me into going. He felt that hunger was "an important issue."

The affair was held at a church and began with brief statements by each of us on who we were, what we did to earn our daily bread, and why we were interested in hunger. I thought about introducing myself as a bank president or a wild animal trainer but ended up telling the truth, that I was an unemployed laborer and that my great-grandmother had starved to death in southern Russia during the 1920's. One other guy introduced himself as an unemployed laborer but no else mentioned any relatives who'd died of hunger.

We saw a slide show and listened to a presentation about infant mortality rates in central Africa and Iceland and sanitary conditions in Asia and Latin America. We were told about the global military budget and the amount that Americans spend on dog food and that the end of hunger as a global issue was "an idea whose time had come."

There was a lunch break after the slide show and everyone sat outside in the golden noon eating fruit and nuts and drinking juice. I walked across the street to a corner market and bought a beer and a bag of pretzels. When I returned I sat down across from the man who'd introduced himself as an unemployed laborer.

He didn't look like a working man and I was curious about him. I asked him what manner of work he was used to doing and learned that he had actually been an assistant something-or-other for a large corporation in a major city. He'd quit his job to become involved with social issues and moved his family to Douglas County, attracted, apparently, by the large number of poor people here. He was making a little money as a handyman and looking for blue-collar employment.

"We've adopted a lifestyle of voluntary poverty," he told me.

"You're kidding!" I gasped. It was as if he'd told me he had chosen voluntary toothache. I was stunned.

I knew, of course, that there was such a thing. The nuns who'd taught me as a child practiced personal poverty in favor of a modest communal wealth, and they told us countless stories about saints who'd given up silks and palaces for hair-shirts and caves. But they, at least, expected a spiritual reward, not insight into an issue or empathy with the oppressed.

I had also read about the Russian intellectuals of the late nineteenth century, who adopted peasant dress and built picturesque little villages to live "the simple life" in. But all their empathy hadn't kept the famines away and later, when Stalin's Five Year Plan brought slavery and death to so many of my relatives, it was those same intellectuals who provided the rationale for genocide.

Sitting there at a picnic table in a sunny chuchyard among smiling Unitarians I realized that I was in a very dangerous crowd.

By the Fall of 1985, there weren't enough of my friends still living in the valley to cover the cost of feeding my cow. I sold her and her calf at auction and waited for the trickle-down.

It was a long time coming, that trickle-down—ten years later we're still waiting for it. The mills and businesses and government agencies recovered but it never arrived for the people who live here. For the first time in our county's boom-or-bust economic history, business boomed while people stayed busted.

Davey was my neighbor's son when we lived in town. He and I worked on the same treeplanting crew for a local mill, during the winter of 1976. In the spring of that year he got a job in the mill. He was 18, fresh out of high school and earning $5.35 an hour.

In the spring of 1986 I ran into him in the grocery store. He'd been out of work for nearly a year but had just landed a job in another mill. He was grateful to have found work, which I could understand since he was now 28, married and a father.

"Well, that's good to hear," I congratulated him. "How much are you making?"

"$4.75 an hour."

I should have kept my mouth shut. But a little mental arithmetic told me that, given the inflation rate, he was earning about half the pay he had as a raw kid.

"Jeeze! You guys are still eligible for food stamps," I blurted.

He looked away, over at the stacked boxes of margarine before he spoke.

"Yeah, well, actually I'm getting more than most of the guys because I've got experience. Starting pay's four dollars an hour."

It was an odd sort of recovery. Employment rates, timber harvest levels and emergency food use rose. The cost of cutting, hauling and milling a million board feet of timber dropped and so did wages and per capita income. Timber harvest levels for Douglas County were 400 million board feet higher in 1986 than in 1978, but produced $55 million less in wages. While the timber industry has become "leaner, more productive and cost-effective . . ." the people have simply become leaner.

In 1989, the seventh straight year of economic recovery brought one in six Umpquans in for emergency food boxes or soup kitchen meals. Local relief agencies estimate that less than 5,000 pounds of emergency food were distributed annually in Douglas County in 1976–79. From 1986–89 the 94,000 people living on the Umpqua ate nearly a million pounds of emergency food every year. None of us saw foresaw that back in 1976, when Davey and I were planting trees, except perhaps, our foreman, Dan.

One morning, toward the end of the season, we drove to work through a bright silvery fog. We loaded up our tree bags in an eerie setting of impenetrable light. The crewmen were a little uneasy and someone cracked a joke: "Hey boys, I think we're in big trouble—I thought I saw Rod Serling standing by the gate when we came through."

"All right men," Dan announced after the laughter died down, "this is it. It's 1976 and the end of the world is here. The Lord's coming for us today, so I want you all out there planting good trees when our time comes."

We laughed, a little uneasily, strapped on our tree bags, shouldered our hoedads and trudged off into the shining haze.

1989

TURNOVER

I read a horrifying statistic the other day which claimed that about half the people in Oregon had been living here less than ten years. At first it seemed improbable. But then I thought of our friends who no longer live here, people we used to see often and visit, whose lives were part of ours. Within a few minutes of ticking them off on my fingers I came up with forty names and stopped, knowing that there were more.

On just one street, the names ran: Bill, Camille, Cindy, Craig, Crystal, Jeri, Jerry, Kelly, Lynn, Midge, Nichole, Sam, Sarah, Sharon, Summer, Tim, Rachel, and Willy.

I thought of potluck picnics and tables crowded with laughing friends who were not yet memories. Gone, some of them living only twenty or thirty miles away, others to more distant cities in Oregon, others to Washington, California, Texas and to parts, and fates, unknown. For me, this town, I realized, is a ghost town, though it is more fully inhabited now than ever before.

Not so long ago, in memory at least (though I find the calendar has added sixteen years since then) I found myself "exploring my options" at the Department of Human Resources Development on Pine Street in Roseburg. There weren't many options. My body was in the first months of what turned out to be years of chronic pain; the local economy was in the first year of what turned out to be five years of double digit unemployment. Nevertheless, job counseling was required by the regulations that governed my unemployment checks.

Sitting in a yellow plastic chair with others who were waiting for the "trickle-down," I practiced Spanish by memorizing the government-issue posters, repeating *"Igualdad de oportunidad es la ley. La ley prohibe discriminacion . . ."* I was expecting humiliation, the one thing I could count on, but secretly hoping that somehow I'd get lucky. Behind the door I would be walking through there might be someone who could show me a secret path leading out of the mucky swamp my life had become.

On the form I'd listed all the kinds of work I'd done, jobs that required the strong back I no longer possessed and which for the foreseeable future wouldn't exist even if I had my health. The bright yellow #2 pencil had been sharpened to a lethal point and a fantasy struck me full blown: I could take a hostage. I saw myself with my arm wrapped around a case worker's neck, pencil poised at his ear, ready to jab it through to his government-issue brain if my demand for a sack of unmarked small bills and a helicopter wasn't met.

He turned out to be she. Sitting behind the desk was a gray skinned, morbidly obese woman. I presented my papers and, in short-of-breath phrases, she asked a half-dozen questions and I gave her my hopeless answers. When she fell silent and studied the form for a few moments I gave up on my pencil project long enough to wonder what would happen if I poured salt on her. Would she dissolve into a greenish foaming mess?

Finally, she fixed her tiny eyes on me and delivered her professional opinion. "Move to Portland," she wheezed, "You should . . . move to Portland."

I protested that I couldn't do that. I explained that I was married, had a child, was a home owner and had no desire to live in a city. I just needed some work that I could do.

She shook her head. "There's nothing . . . for you here," she gasped, "Move to Portland . . . think about it."

⚜

I have thought about it many times over the years—not about moving, but about her. Whenever I hear of another neighbor moving for lack of work, abandoning our valley for the urban wilderness, I see that woman and hear the sucking sound of her breath and the whispery advice she must have been handing out every ten minutes, day after day all through those hard times when forty percent of the workers in this county left their homes in search of something they could do.

This is a tough place to live, whether you've newly arrived or been here for decades. Some people manage to get by, and a few even prosper, but most of the folks who come here to settle in, move on. They arrive, fresh from the cities, all starry-eyed and infatuated, feeling like they've snuck through the back fence of Eden. It takes a while to realize that here too life has a cost. Everyone, eventually, begins to suspect that their lives might be more comfortable if they lived somewhere else. Accepting the necessary struggle to live here is much like getting married, "for better or for worse, for richer or poorer, in sickness and in health, 'til death do us part."

Our valley's chief export is not lumber, minerals nor cattle but, people, particularly the young. Nearly every family which survived the hard times despite staying here, has at least one child who, once on their own, left this small town for the city. I watch them go and hope that perhaps they'll return eventually, better for the experience. Once in a while one of them does come back here to live, but not often enough.

There is a pull to cities, a sort of gravitational well like a black hole—the pull of money. It reaches into the farthest corners of the world, sucking the life out of rural communities through the horrors of environmental degradation, cultural impoverishment and population loss. It seems as inevitable as the seasons, as though it were due to some unquestionable law of nature, like water flowing downhill and carrying everything that falls into it out to sea.

Leafing through analytic reports, heavy with jargon and charts and projections, we find no trace of good nor evil—only digitalized fragments. It is as though our lives were not lived as human lives, whole and in relation to unfathomable mystery, but as isolated segments which generate numbers. Only these numbers—pure, austere, measurable, comparable and unarguable—seem to matter. Only the numbers are particular, unique and individual in relation to time and geography. In the glowing, futuristic world of government planning, social engineering and economic development, everyone, everything and every place is interchangeable.

And yet, we continue to live our particular lives anyway, and we live them in our particular places. We live not by computer models, spreadsheets, diagrams and graphs, but by the complicated, demanding and necessary daily social juggling act imposed upon us by human nature and Mother Nature.

Only through this daily exercise can we come to useful knowledge. To know our neighbors requires a good long while. To know ourselves is the work of a lifetime. To know our home lands is the work of generations. To pack up and leave is to abandon the faith, the hope, and the charity which ensure survival. Who will care for this place and the people who live here, when nobody is left who understands this land, these people?

A few years ago I met a man who'd spent the first twenty years of his adulthood living far away from the place where his cradle stood, first at college and then working for an environmental organization. He wanted to save the world from destruction and sacrificed much so that he could be free to do so. His work took him to Washington, D.C. where his organization had its offices, in order to lobby the federal government. There he lived out of a series of apartments while his organization sent him to remote villages in Asia, Africa, and South America to learn of the environmental and economic problems the native people faced.

He wrote reports on all those places and the people there and presented them at conferences in the world's centers of power—Washington, Tokyo, Stockholm, London, Rio de Janiero. It was heady work, exciting and challenging. He and his colleagues drafted proposals that helped shape international agreements on trade, technology, pollution and the fate of endangered species of plants, fish and animals

One day, in a remote jungle village, an old woman asked him to describe his home. It was a simple question but one he found he didn't know how to answer. What should he tell her? About his childhood home town, a place he'd left long ago and occasionally visited? About the latest set of rooms in a large building where he kept his goods and where he slept when he was between journeys? About the large city where he worked and which he simply endured as an inconvenient annoyance? He felt suddenly his own poverty and ignorance and was ashamed because he had no home which he knew intimately and cared deeply about.

Like many people, he had no allegiance to his daily world. He lived for a future world, one which he earnestly hoped and prayed would come about. He neglected his neighbors and knew only his colleagues. He had searched for metaphorical "common ground" but never really considered the ground beneath his feet. He could speak to large audiences about global issues but the simplest sidewalk conversation left him feeling inept, awkward and embarrassed. He'd dedicated his life to saving the planet but had never concerned himself about what was happening in the places where he lived.

The revelation, he told me, had a profound effect on his life. He'd discovered a concept which he called "community" and had returned to his home town to live. He'd written a book on this issue and founded an organization to promote "a sense of place."

I wished him good luck.

1997

# THE POLITICS AND PREJUDICE

## OF OLD GROWTH

*"There are many fine things that we cannot
say if we have to shout."*

— HENRY DAVID THOREAU, WALDEN

Fear has a way of sneaking up on you. It's an odd, unsettling moment when you realize that you've been afraid for a long time and have been trying to hide it from yourself.

For me, the moment came on a sunny August morning in 1988 when I stood on Main Street in Canyonville, Oregon, a small town near my home. I had brought my son and his friend to the annual Pioneer Days parade. The boys were having fun, dashing out into the road to scoop up candy tossed from the passing floats and old cars. My niece marched by decked out in calico and a sun bonnet. My nephew strode past, banging the bass drum in the Days Creek Marching Band.

I always feel comfortable and expansive at small town parades, like Walt Whitman in the crowd, everybody's kindly uncle, soaking in the joyful sounds, the gaudy colors and the cheerful faces of my family and neighbors.

The parade had something new that year though. One of the floats was draped in yellow ribbons and featured a logger in hickory shirt, red suspenders and hard hat wielding a chainsaw. He mimed the act of falling a fir tree in the top of which was tied a tree-sitting hippie doll. The tree leaned precariously toward the truck bed which carried a banner asking "Who Will Hit The Earth First?"

My heart sank as my neighbors laughed. I realized that the doll was a caricature of myself and many of my friends, long-haired and bearded. I imagined the horror of actually being up in one of those groaning giants in flight. Having worked as a timber faller, I knew the tension that comes when cutting down one of the big old trees. I heard again the cracking of wood during the back cut, the creaking and whooshing of the tree in flight and the apocalyptic roar that tons of wood striking the ground makes.

Of course, I understood it all. It would be difficult to find a Northwest county more dependent on the timber industry than ours was. I knew that my neighbors were frightened, running scared in the face of an uncertain future. But now I found that anyone who even looked like they might disagree with the local majority was bearing the blame. Simply by my appearance, I had become an outsider in my own community.

It occurred to me later that these timber workers and their families were wonderfully good-hearted people who'd been inexplicably maligned. I realized that the float was not really an expression of hatred so much as a derisive reaction to a public debate in which the proud title of "logger" had become a pejorative term. It was a sort of perverse acceptance, through parody, of the role of heinous men with chainsaws which they'd found themselves saddled with. Our slanders always say more about ourselves than about our enemies and, all too often, shapes their character to fit our fears.

Of course, I don't fear my neighbors, the people I know and who know me. To know, to really know, someone is, an act of love, requiring patience and an open mind. But what about those who only think they know me, who have fixed notions about categories of people—notions which keep them from seeing individuals?

It's much easier to hate (or love) a class of people than a particular person. People are complex. There's always something about an individual that gets in the way of our prejudice. "Am I not vast enough to contain contradiction?" asked Walt Whitman.

Aren't we all?

I, too, yearn sometimes for pat answers and the reassurance of dogma. I almost wish, sometimes, that I could be a bigot. It takes me a lot of work to really know people. I have to listen to them and watch them and learn about their past, their hopes and fears, their lives. It's terribly time consuming. It is also heart rending.

⁕

Much of the frustration which environmental and industrial activists feel stems from the failure of confrontational politics to resolve the forest management dispute. Traditional political debate is limited in what it can address, working well with specific issues that can be decided with a simple vote. Unfortunately, the system, with its short-term, piecemeal approach, isn't able to resolve complex, long-term, broad, fundamental problems. The social, economic and environmental crisis brought on by forest management is at heart a cultural and scientific problem, not a political one.

A debate between government agencies, political organizations and corporations is bound to lack a human dimension. Institutions simply don't have hearts and by their nature discourage compassion.

⁕

In the spring of 1986, two years before that memorable parade, I helped organize and run a bioregional conference in Ashland, Oregon. Four-hundred political activists gathered to take an "holistic approach" to dealing with social, economic and environmental problems in southern Oregon and northern California. For well over one-hundred years the people in this region have called it the "State of Jefferson," in recognition of its uniqueness. But for the occasion it had been dubbed "The Klamath-Siskiyou Bioregion" which had an aura of sociological legitimacy that the traditional name lacked.

Although there was a great deal of talk about seeing our region as a whole, crossing class lines and working democratically for the benefit of everyone, it became obvious that "we the people" really meant "we the people who think alike" rather than "we the people who live here."

A workshop on "Long-term Environmental Activism" turned out to be a group of environmentalists talking about the dangers of learning to like some of the government and timber industry officials they were forced into contact with. It was agreed that familiarity with their opponents carried a terrible risk of being "co-opted" into accepting a compromise. For an hour, these earnest, well-intentioned activists searched for ways to maintain purity of purpose in the face of friendliness without ever mentioning compassion.

Failing to take the true measure of opponents is an all-too-common failing among political activists. It has brought sudden disaster to many movements and opened the way for unintended harms despite the best intentions and most noble aims. The problem is that no matter how exact the measurement, the end result is only an approximation.

Faced with inevitable imperfection and the impossibility of actual uniformity, craftsmen speak of tolerances. If a measurement falls within a certain range it's "close enough" because no matter how similar two of a kind may be, in the end there's really only one of each.

But somehow, although we know this on the physical level, we don't seem to understand that this is true of people too: there's only one of each. People often insist on a lack of tolerance in dealing with each other.

We have, in measuring people, words to work with. But words, particularly in political discussions, are very unreliable measuring tools. Unless we constantly test them against what we actually see occurring, we run the risk of fooling ourselves with notions that have an internal logic that doesn't match up with reality.

Words are not only the tools of my trade but the raw material as well. I love the language, yet I distrust it—as any artisan who respects both the work and the material must. I've spent too much time practicing artifice to put much faith in words.

I once saw a magician sitting at a table alone between performances, playing solitaire with a deck of cards. I felt an immediate kinship with him because my own years of practicing the techniques of persuasion have left me as lonely as he. In his case, it was society's distrust of his dexterity that left him unable to join a card game; in my case, it is my skill at of sleight-of-mind which leaves me unable to trust the players in any political game.

The stereotypes of "preservationists" and "timber barons" have just enough truth in them to reinforce the images. Some mill owners really are greedy, some mill workers and loggers are truly ignorant and brutal, some environmentalists are, in fact, utterly insensitive to the needs and outlooks of blue-collar workers. But these individuals are actually rare. The few truly narrow-minded ones I've met are simply pathetic people whose dismal lives lead them to seek self-worth through political activism. Industrial and environmental extremists are actually much more similar to each other than they are to the moderates within their own camps.

Chaung Tsu, the Taoist philosopher, wrote, "When wrestlers pit their strength against each other, they begin in a lighthearted, open frame of mind but they usually end up looking angry. At the height of the contest, many crafty tricks are played." When push comes to shove, the farther apart two sides grow the more they become the same.

During the winter of 1979–80 I worked on a tree planting crew. I became friends with Steve Hoeffler, a planter from Monterey, California, who had been working for the Forest Service here in Oregon for a few summers and planting trees during the winter.

One day at lunch, Steve told me about his younger brother, who lived in Santa Cruz, California. His brother, inspired by Edward Abbey's *The Monkeywrench*

*Gang,* had been vandalizing logging equipment on the weekends.

I thought about Stewart Clason, an old gyppo logger who'd taught me timber falling and logging. I imagined him coming to work in the morning to find his geriatric equipment vandalized.

"Jeezus, what a jerk," I said.

"Yeah, I told him to cut that shit out. It's stupid," he sighed, "But he thinks he's some kind of warrior or something. I tried to tell him that if he really wants to change things, he ought to get a job setting chokers and get to know the guy and then talk to him, instead of sneaking around and messing up his show. The dumb shit's never worn a pair of caulk boots in his life."

Three years later, Steve ran a garden hose from his pickup's exhaust pipe into the cab, leaving a bright red corpse for the sheriff's department to investigate. He was working for the U.S. Forest Service at the time, in the timber sales department at Toketee Ranger Station on the Umpqua National Forest. I went upriver to visit him about a week before he killed himself. He'd broken up with his girlfriend three weeks before and was camping out in the woods by himself.

He didn't talk about the breakup much, mostly about his gloomy outlook on the environmental health of the planet in general and the over-cutting of old growth on the Umpqua National Forest in particular, and of how tired he was getting of being put down at work by the rest of his timber sales crew for being "that goddamn environmentalist." He told me that he'd been very lonely, living like that.

Based on ignorance, born in fear, driven by tension and frustration and nurtured by demagoguery, prejudice follows a five-step path of escalating actions. The first step is slander which gives it a voice. Then comes avoidance of the hated group, making the lies easier to believe. Next, discrimination, because, after all, something ought to be done about those people. Personal attacks follow when discrimination seems too subtle. If isolated attacks won't make them leave, there's always the ultimate solution—genocide.

Though the forest management controversy never got beyond the fourth level around here (and even that in only a very few instances) each step lays the foundation for the next, making it possible. Not every dog that barks will bite, but no dog bites without barking first.

In Douglas County, Oregon, I first heard the barking in 1978, during a federal review of potential wilderness areas called RARE II, for Roadless Areas Re-Evaluation, second round.

I was vaguely aware of the controversy and generally in favor of setting aside some of the forest, though I hadn't thought about it much, being busy working fifty hours a week at a local mill. My wife was pregnant and we'd just bought a small place, so my worries were about maintaining my health insurance and making my land payments rather than the health of the forest.

For two months that summer we were urged to oppose wilderness. The foreman shut down the machinery one day to lecture us about the threat that "those goddamn environmentalists" posed to our livelihood and offered a free steak dinner for the entire crew if our shift filled out more public comment postcards than any other crew.

I looked at mine, saw that whatever I wrote would be open to my boss's inspection, and tossed it into the chipper feed belt when no one was looking. I wasn't concerned about the politics of forest management but, as a worker, I was offended by the postcard.

About a month later, I was sent up to the millwrights' shop to fetch a gear for one of the machines. When I got there I found the head millwright busily fabricating hatchets, maces and clubs out of scrap steel. The plant foreman was standing there with him and they were laughing excitedly. I learned that there was to be a public hearing that night at the county fairgrounds on the RARE II review. "We're gonna go out to the fairgrounds and kick-ass on them environmentalists," the millwright told me. Nobody was attacked that night and a week later the weapons were lying in the scrap metal bin without having drawn any blood.

I've often wondered about the millwright and his makeshift weapons. Perhaps he simply thought better of it and left them at the mill. It might be that he and some others actually did bring them to the meeting but changed their minds when confronted with the opportunity to use them. Or maybe they just couldn't pick out anyone to attack because you really can't tell, just by looking, what someone's politics are.

1989–1995

# TALKING TRASH

"*The Revolution was effected before the war
commenced. The Revolution was in the hearts
and minds of the people . . . This radical
change in the principles, opinions, sentiments
and affections of the people, was the real
American Revolution.*"
—JOHN ADAMS, LETTER TO HEZEKIAH
NILES, FEBRUARY 13, 1818

I was up at the dump a while
back—the same place that I've been hauling our household trash to for twenty-two years now—and I got to
thinking about the changes our little Southern Oregon
timber town has gone through over the years. There
used to be a hole in the ground there where we tossed
everything we discarded—tires, paint, used motor oil,
furniture, animal carcasses, garbage, plutonium. Every
once in awhile someone would set it on fire and the
heap would get smaller for a while. Maybe twice a year
the county sent a man with a bulldozer down to com-
pact the mess and spread some dirt around.

Nowadays we have what's called a "transfer site" which sounds different but smells pretty much the same. We throw our stuff into a metal dumpster which gets picked up by a semi-truck and hauled downriver to the county's "sanitary landfill" where it gets dumped into a big hole in the ground and a man on a bulldozer works five days a week compacting the mess and spreading dirt around.

I guess that doesn't sound like much of an improvement but things really have changed. We can sort our trash now, recycling paper, tin, glass, plastic, appliances, motor oil, leaves and grass clippings. This saves us room at the big dump and makes a little money for the local charity that sells what we sort out.

While I was musing, instead of tossing, one of my neighbors pulled in. He was a logger, a timber faller in fact, judging by the chainsaws, oil and gas jugs, axes and road warning signs in the bed of his crew-cab pick-up.

Our county calls itself "The Timber Capital of the Nation" (which isn't too far from the truth) so loggers are a common sight in these parts. The bumper-sticker on his truck read, "Help Ruin America—Join an Environmental Group" which is a pretty common one now, like the ones that say "Keep Oregon Green—Stop Clear-cutting." Twenty-two years ago you never saw anything like that around here.

Well, the first thing he did is what just about everyone does nowadays. He pulled up by the recycling shed and dropped off his newspaper, glass, tin cans and plastic milk jugs in their appointed bins. It was all so commonplace that, if it hadn't been for that bumper-sticker I never would have noticed.

My first reaction was delight. I'm not a big fan of bumper-stickers and the sight of someone seemingly contradicting his own slogan left me with a smug satisfaction. It occurred to me that my neighbor, despite his evident politics, had been seduced by a cultural change. Two decades back, recycling was for hairy-legged granola women and subversive simple-lifers—now it's mainstream, something we all do because it's what we feel we ought to be doing.

The smugness lasted until it occurred to me that the "new" transfer site with its recycling bins was ten years old. Where the hell had I been all that time? How could I have not noticed that my neighbors had accepted this once-radical change, endorsed it whole-heartedly as an act of common decency? Mores, the sociologists call this, the agreed-upon ways of doing things which set the tone for the entire community.

My job, as an artist, is to keep my eyes open and notice what's going on around me. Here was something both subtle and significant that had been going on all around me every week for ten years—520 or more trips to the dump and I'd never caught on to a trend that took place right in front of my own nose.

I took comfort in the knowledge that I'm not alone in missing out on much of the true complexity of small town rural life. In a way, the logger who recycles could serve as a rural/urban and blue collar/professional class litmus test for prejudice. Does a timber worker have Green nature? Whether you answer "yes" or "no" or "maybe" or "I don't know" says a great deal more about you and your role in our culture than it does about the situation itself.

—✦—

My poet buddy of mine, who is an ex-logger and ex-environmental activist, likes to tell an anecdote about the time he was blowing down Interstate 5 along with the director of an environmental activist group. His passenger came suddenly unglued as they passed a log truck loaded with big old-growth #1 Peelers.

"You motherfucker!" the director shouted while flipping the driver the bird, "We're going to shut you down, you tree-killing sonovabitch!"

Maybe the truck had a bumper-sticker that he found offensive or maybe it was just the sight of those fat logs that set the director off and turned some poor schmuck of a truck driver into a scapegoat for all that was wrong and frustrating in the director's pitifully small world. Yet many log-truck drivers also "reduce, re-use and recycle" and the target of that up-thrust finger may have been stacking his newsprint in a recycling bin for years. Maybe the director's contorted face and his one-finger salute confirmed the trucker's worst suspicions about environmentalists and their alleged, "socialist agenda." In all likelihood, the driver never even noticed.

My neighbors find it unremarkable that mill workers, loggers and log-truck drivers recycle. To them, anyone who thinks that timber workers don't love the land is just another ignorant (and arrogant) outsider trying to tell them how to live without bothering to first learn about their lives.

~~~

Another friend, a tie-dyed-in-the-wool middle-aged flower child, runs a health food store in a local small (population: 1,100) town. A half-dozen years ago, when spotted-owl fever was running about 109.7 degrees Fahrenheit, he was "encouraged" to place a dayglo-green placard in his store window reading: "This Business Supported by Timber $$$$."

"Screw the mill owners," he told the committee, "They've been shafting everyone around here for years—running the gyppos out, cutting wages, killing the fish. Now they got their tits in a wringer and they want me to help get them loose?"

As the only business on Main Street that didn't display the placard, he found himself isolated and boycotted. During those days, two Earth First!ers from Santa Cruz showed up in his store. No doubt feeling more comfortable in the store than they'd been out on the sidewalks, they started loudly voicing their views on the old-growth question, blaming it on "asshole redneck loggers" who'd been "duped by the timber-nazis."

"You guys don't know shit," my friend told them, "You're the only assholes around here. Quit bad-mouthing my neighbors and get the fuck out my store."

---

I gave up political activism years ago—at least on being a member of any group with political aims—though I still participate as a sort of freelancer, attending meetings and rallies when a cause is important enough to me and voting at every opportunity. In part, my decision to drop-out from that scene was simply due to a realization that I'm no good at it. My distrust of any group too large to sit together around a picnic table and of any ideology more specific than "for the general welfare" makes me a disruptive element in any organization.

Besides, I'm always haunted by the notion that I'm too ignorant to tell others how to live their lives, and always horrified at the chutzpah of those who claim to know what's right for everyone. I've spent too many years questioning myself to accept the pronouncements of others without first turning Michel de Montaigne's humble motto, "What do I know?" inside out by asking, "What do *you* know?"

Polemics, manifestos, ideology and the degeneracy of political language seem hopelessly simple-minded, arrogant and deceptive to me. Worse yet, reading it bores me to tears—the only unforgivable sin on the part of any writer expecting to earn a paycheck and an audience. The best writing, the kind that achieves very high levels of artistic merit, doesn't seek to provide answers and to persuade—it raises questions, makes us doubt, hints that there's always more to any situation than we can possibly understand, challenges us to draw our own conclusions.

Politics, with its short-term, confrontational focus on winning and losing particular immediate battles, can't help but generate more trash-talk than sensible words. Though both Mother Nature and human nature (which is just an aspect of the former) are delightfully complex and therefore endlessly interesting, it is hard to build a mass political movement founded on complexity and moral ambiguity.

The truly important stuff just won't fit on a bumper-sticker—at their best, even a book-length collection of essays or a novel can only sketch the barest outlines. Still, there is a role for art in resolving social, economic, environmental and political disputes.

We often talk of art in terms of "Culture" (with a capital "C") as if it were important for its own sake—*Ars gratia artis*. But the ultimate purpose of art is to enhance our chances of survival, to, in some way, change "life as we know it" for the better. It is culture, "little `c'" culture, the things we do every day, which gives meaning and purpose to the arts—not the other way around.

Politics, too, is an aspect of culture. The political discourses of today are a result of the cultural changes of the past. It is only after people change their daily habits of thinking and ways of doing that the political debate on the consequences of those changes begins. Lately, I've begun to use the term "cultural activism" to describe what I and others do in our work as literary, visual and performing artists.

Back in 1989, just before I gave up on politics, I spent a weekend at the county fairgrounds sitting in a booth at the Spring Fair, an annual crafts show. It was an informational booth though, so we weren't selling things—just giving away facts and ideas. You wouldn't think it, but at the time it was a dangerous job—several people whom I knew had been getting death threats for less than that.

The spotted-owl controversy was going on and many of my neighbors were understandably upset about maybe losing their jobs and homes and their way of life in order to protect a bird nobody had ever even seen. It didn't make sense to them.

"Yellow Ribbon Fever" we called it, because of the plastic yellow ribbons the mill owners passed out all over the county so that folks could tie them to the antennas on their cars and pick-ups showing the world that, if it came down to it, they'd rather see the bird go than lose their livelihoods.

Of course, the reality of what was going on was much more complicated than that—in fact it was, and still is, one of the most complicated problems I've ever seen. But "jobs versus owls" is what nearly everyone, left, right and center, seemed to think it was about at the time. (Actually, as slogans go, "jobs versus owls" was much more effective than, say, "Subvert the Dominant Paradigm"—less abstract, fewer syllables, something you could sink your teeth into.) Having been repeated enough, it was taken not just for truth, but as the whole and single truth.

So, there we were, the board of directors for Umpqua Watersheds, our little local environmental activist organization, right out there in front of God and everybody, trying to explain things to our neighbors and wondering who'd be the next to get a midnight phone call, or lose a job or have a car vandalized for expressing our opinions. But mostly, we listened more than we talked.

The funny thing is, it didn't matter much which side people were on. Just about everybody who stopped by started out by talking about "them," the other side. To hear the litany of complaints and accusations you'd think that there was a war on and that unconditional surrender was the only way to end it.

Well, there's no use arguing with people when they're upset. We just let them blow off steam until they settled down. Then we asked them all the same question, "Forget about 'them' for a minute, what do *you* want?"

It kind of stunned them for a moment, as if they'd never been asked that before. Their faces changed from indignation to shock and confusion, followed by puzzlement and distrust, and finally, resignation and humility. In the end, they'd let loose with a what-the-hell sigh, take a look around and lower their voices, "Well, it's pretty simple really . . ."

It turned out that nobody wanted to fight; nobody wanted to harm the land or the critters; nobody wanted people to lose their jobs. Everyone was certain that there must be a better way of doing things.

"But what can we do?" they all asked, "What can we do?"

1996

It was not really surprising but, well, disappointing to hear that I'd been called an "eco-terrorist" by one of my fellow Umpquans. The news was second-hand, of course, which somehow made it worse. Whoever pronounced the judgment, whether she or he, hadn't bothered to tell me about it, but let it slip, off-hand, as if it were a well-known fact.

"Him? He's an eco-terrorist," as though there was nothing more that need be said, as though I had it printed on my business card, "Bob Heilman, Eco-terrorist."

I did what people usually do when confronted with something odd and shadowy and disturbing. I feigned indifference, laughed and pretended that it amused me. After all, what else can you do in such a situation? It was merely an anonymous slander, not meant for my ears anyway and no doubt committed by someone whose life is probably a good deal sadder and more wretched than my own.

Still, I resented the accusation. I began foolishly marshalling my arguments, standing before an imaginary judge and jury, reviewing twenty years-worth of statements public and private to see if there was anything I'd ever said or done that would lead a reasonable person to conclude that I am a menace to my neighbors.

At first, I focused on the first half of the odd compound word, "eco," as in "ecology." Yes, I'd spoken and written about environmental matters. I can't help it, having lived in Douglas County, Oregon during a time of harsh words and desperate acts centering around what are known as "forest management practices."

Generally I've come down on the side of the soil, water, sunlight and air. But I've always been of two or three minds on this and nearly every other matter. Conflict, I believe, is unavoidable in a free society; compromise, I believe, is essential to preserving it. Complexity is what I find wherever I look, whether at the natural systems which provide us with life or the culture and politics with which we deal with nature and each other. I suppose it's a natural approach for me, a character trait that led me to writing and one that has been reinforced by the craft itself, with its necessary habit of carefully observing and then describing what I see.

Terror, then, the second half of the compound may explain this. Although I have always opposed extremism and its violent expression, terrorism, yet, somehow, someone is afraid of me, frightened enough to denounce me as a terrorist. Why me? Why anybody?

I cannot say, except that it is human nature to try to put a face to your fear. Still, it's an unsettling experience, sad and a little scary to know that for someone in my community my face is the face of what they fear.

Poor faceless one, I wish I could help you to face your fears. I too know what it is to be frightened. I would try to comfort you, if only I knew you and you knew me.

***

Someone has girdled a sugar pine tree upriver from my house. The tree is probably 500 to 600 years old, a century or two old back when Shakespeare was a snot-nosed toddler, the second tallest of its species in the world, 265 feet tall and 7 1/2 feet in diameter. It's been something of a local attraction for decades. Tourists stop by to admire it and generations of local school children have measured its circumference by holding hands to form a circle around it. The Forest Service is optimistic. They believe that the tree will probably survive since the life-giving cambium layer was not completely severed through the full circle. They've sealed the cut and are waiting to see whether the tree will still be alive five or six years from now.

No one knows who tried to kill the tree or why. Perhaps it was simple vandalism, some thoughtless act of more-or-less random violence. People go into the woods for many reasons—sometimes to find peace in solitude, sometimes to loose their rage in secrecy.

Perhaps the motive was revenge, personal or political. The US Forest Service and the federal government have plenty of ill-wishers. No one has come forward to claim responsibility, but there are whacked-out people out on the fringes of America, some of whom suffer from frustration and anger and some from genuinely sociopathic or psychotic obsessions. It's impossible to say who or why, and so, there are only suspicions.

"A thing of beauty is a joy . . ." well, unfortunately, not "forever," but only until some yahoo destroys it.

Among my dreams is a recurring nightmare, not a frequent one, but it comes to me often enough that I recognize it as having visited me several times over the years. I find myself traveling about in my home valley. I know it is home because it feels like home, yet it is a landscape which has been altered beyond recognition. The mountain slopes are ravaged, bleeding red strips of soil, boulders, and logging debris. The streams are clogged with mud and slash. I drive an old pick-up along a logging road, surveying the damage. The road keeps getting worse and my heart grows heavier. The road becomes a pair of uncertain ruts climbing more and more steeply toward someplace I cannot see. My body presses into the truck's seat as it approaches vertical. I gun the engine, the tires clawing while I hold onto the steering wheel, worried that the poor truck will spill over onto its back like some hapless metal tortoise. The truck keeps going, the road keeps getting worse, the destruction on either side more awful.

Erosion, the slow degradation of our natural and cultural habitat, is a reasonable, and pervasive, cause for fear. Most of us suspect, at times, that we are hurtling through the devastated landscape of what was once familiar, driving a doubtful vehicle on a rocky road to a future that can only be worse. The knowledge of mortality, personal, societal and ecological, can be a source of compassion in our lives. It can also leave us feeling isolated, frustrated and angry.

When fear grows, the temptation grows to embrace ideology. A rigid and simplistic systematic approach offers the comfort of relief from doubt and the promise of a ready-made answer to every troubling question. We fear moral ambiguity with its obligation to admit that we don't always have a simple answer, that we might be wrong, that, many times, the choices are not clearly right or wrong but often the hard, sad choice between tragedies.

I have met perhaps a dozen genuine extremists over the years, unhappy, inflexible people who force everything they see through the distorting lens of their rigid beliefs. Every single one of them has had a life history of failure, defeat, frustration and anger which preceded their fanaticism. They have all been losers, the lonely, the alienated, the sufferers of loss in love and in money. Their extremism provides them with a sense of being ennobled losers, the hapless victims of sinister forces rather than of their own mistakes. They are fond of elaborate conspiracy theories and they invariably attribute much more cleverness, power and control to those they perceive as their oppressors than any person, or group of people, could possibly possess. They are the followers, the joiners of cults and causes. Within the shelter of a mass movement they shed their personal past like shabby clothes and don the glittering theatrical attire of martyrs and heroes.

The psychological basis of fanaticism is well known. Extremists operate on a principal known as "projection," a bit of sleight of mind in which our inner turmoil finds a target outside ourselves. We detest most in others what we fear most in ourselves. To face our own fears—of inadequacy, of guilt, of powerlessness, of accepting responsibility for our own actions—is never pleasant and always difficult. It would be easier to understand that we are caught in the leg-hold trap of fear and must either die or gnaw our foot off, if only we didn't hear so many deceptive voices offering a simplistic and seemingly painless solution.

People ought not to make other people feel small and powerless and afraid. "Mean People Suck," a popular bumper sticker reads. Another sign of the times crudely demands, "Don't Be a Dick." By and large, most people, most of the time, don't need to be reminded of the need for simple decency. People know how to treat other people. The Golden Rule applies everywhere and is known to virtually everyone. Outside of a small percentage of seriously stunted individuals no one disputes the propriety of granting each other a measure of forbearance.

For the most part, "the evil that men do" is petty evil and we are all, at times, guilty of failing to honor another's humanity. But the difference between mass murder and rudeness is merely a matter of degree and intensity. Underlying each is an overblown sense of self-importance and a corresponding disregard for others. We are free to choose between fear and love whenever we deal with ourselves or others. Too often, we fail to choose wisely.

2003

SMALL TOWNS AND QUIET VOICES

I t was just a deal that went sour, like so many other schemes that I've hatched over the years. But some things that should have been said, and could have been said, and would have been said, weren't said because of something I said.

It's not as if I made an innocent blunder or got blind-sided by forces beyond my reckoning. I had a hunch it would kill the project and went ahead and took the risk. And who knows, maybe it wouldn't have worked out for other reasons, though it sure seemed like a done deal at the time.

To understand what happened and why requires understanding a people and their place: Douglas County, Oregon, "The Timber Capital of the Nation," my home. It's a large rural county, 5,000 square miles of forested mountains, with a small, scattered population of 94,000 people. It has a reputation, when people bother to think of it at all, of being a redneck cultural backwater, the home of hillbillies, crackpot secessionists, and Holy Roller revivalism.

While there is some truth to that stereotype, the reality is more complex, as is always the case with stereotypes. Although many of my neighbors, perhaps most, hold views on generalized issues that urban liberals would find appalling, when it comes to the specifics of day-to-day living they are traditionally sweet, honest and reasonably tolerant people.

Unfortunately, the old-growth timber harvest controversy created some serious problems here. As a citizen and as a writer I tried to help people understand this complex problem and its ramifications. But facts and reason, I learned, are not politically expedient.

The industrial and environmental outlooks each spawned mass movements with opposing views and neither group was interested in presenting complexity, which might raise doubts in their followers' minds.

"We face a long, uphill battle against a relentless foe whose arrogance, lust for power and disregard for human suffering seem to know no bounds," one pamphleteer wrote regarding the opposition, a statement that could as easily have come from one camp as the other.

The words became flesh and, first through threats and then by covert actions, people were getting hurt. Some store owners complained that a "green list" urging customers to boycott their places was circulating. Four of my friends, three environmentalists and an industrial activist, received death threats. Another lost his job.

The debate took on the form of harassing phone calls, midnight beer bottles smashed on driveways, a broken windshield, and a note asking "Who's watching your wife and kids while your [sic] at work?" It was a hidden thing, a painful festering abscess, talked about in private but never mentioned publicly.

I looked for a subtle way to help heal the wound. It seemed to me that we Umpquans were starting to see ourselves and each other as outsiders saw us, as an issue rather than as a people. I was doing some radio and television commercial voice-over work for a small advertising agency and I wanted to start up a series of short local color sketches which would focus on the things that make us glad we live here.

The idea was to talk about the little things we held in common as a people, the things we all cared about and which made up our lives—the landscape, the seasonal cycles of crops and weather, the wild creatures living in our forests, rivers and fields, and neighborliness. Maybe if we could talk about our everyday lives instead of our fears about the future we'd remember who we are. I wanted to send a flock of sparrows to peck away at a boulder of intolerance.

The agency was enthusiastic and agreed to try selling the project. They were sure we could get some local business to sponsor it as a radio feature and I hoped to syndicate a weekly column to local newspapers. I spent two months writing short word-sketches to be called *Upriver Reflections*.

I thought of the project as a series of small hand-painted tiles, each a picture in its own right, forming a large mosaic mural. It was exciting work, and trying to get the tone just right—uplifting but not preaching, lovingly without being schmaltzy, artful but not dishonest—was challenging and satisfying.

Two months into the project, I was laying the words down, building up material, there were leads on sponsors, a demo tape was planned and, *mirabile dictu!*, it seemed I would earn a modest steady income while doing something both commercially and socially useful.

But then *The Oregonian* sent some reporters down from Portland for an in depth series of articles on how the timber controversy was affecting us. It didn't take them long to hear about all the harassment and soon dozens of people knew that the ugly secret was no longer going to be hidden. The local daily, *The News-Review*, got wind of their inquisitive competitors' plans and, a week before *The Oregonian* published its series, did a front page piece exposing what everyone already knew.

I felt relieved, as if a painful boil had finally been lanced. I waited for the public reaction, for an editorial or a public comment from someone suggesting that things had gotten out of hand. But not one local journalist, politician, preacher, educator or activist spoke up. The only public reaction was a single letter to the editor saying that one of the victims was a traitor who deserved whatever he got.

Since no one spoke out on behalf of the victims, I wrote a guest editorial for the local daily pointing out that intimidation in the name of politics is shameful and stupid as well, because preserving the right to express opinions is more important than the outcome of any passing issue.

Nobody publicly agreed or disagreed with me, though several people stopped me on the street or called me on the phone or wrote me letters to thank me for having voiced their own concerns. I didn't get a pipe-bomb in my mailbox but I did get a letter from the advertising agency warning me that the *Upriver Reflections* project might have to be dropped if I continued to make controversial public statements. As it turned out, the one piece was enough to kill the project.

I understood. It's a small place where everyone knows everyone. In a polarized situation, a moderate stance can be highly controversial and why should a business take risks? It wasn't really censorship or black-listing—I could have pushed it on my own, if I'd had the heart. But still, I wonder what effect the project would have had on us all. I wonder what other calm and quiet things haven't been said or have been drowned out in all the shouting.

1992

# THE ENEMY AMONG US OR
## THE ENEMY WITHIN US?

Two years ago, I was at Shasta Lake, waiting on a dock with some time to kill when I met a woman from Klamath Falls. Since we were both away from home, our conversation soon turned to Oregon and the old growth timber debate between the environmentalists and the industrialists.

"I don't know what to think," she admitted, "I have friends who work in the woods and I'd hate to see them lose their jobs but I'm worried that the forests really are being cut too fast too. It's terrible. I keep thinking that there must be something wrong with me."

"Well, that's normal really." I told her, "It's a tough situation with no easy answers. You should be confused and worried because it's a real gut-wrencher. That's healthy. It's the people who don't feel that way who have something wrong with them. Complex problems don't have simple causes or simple solutions. Life just ain't that easy."

Sometimes the ability to see both sides of an issue can be a curse, turning our own hearts into disputed territory, just as the community itself becomes divided. We sometimes yearn for an end to conflicting notions, something comforting that we can hang on to which settles the matter. But this uneasiness, as hard as it is to live with, is unavoidable in a democracy such as ours.

One easy way out of the dilemma is to accept wholly one version or the other of the conflicting views. This simplifies things by reducing matters to two sides instead of many different views. It also brings the comfort of belonging to a group of people who all feel the same way. When an issue, any issue, stirs us up, we can join with others who have a ready-made theory (usually a conspiracy theory aimed at anyone opposed to the group) and escape the discomfort of open mindedness.

Of course, accepting an "us against them" view doesn't really end our frustration at all. It just gives us a convenient target for it. Reality is a tough and bitter pill to swallow, but sugar coated lies go down easy. The problem is that reality lingers on after the sugar high fades.

Having accepted "arrogant radical preservation- ists" or "greedy timber barons" as the source of the

conflict we must spend a great deal of energy defending
our group's theory against the incursions of reality. The
all-too-human desire for an answer we can believe in
leads to frustration when the answer we choose is too
simple to fit what is actually happening.

Conspiracy theories ignore some basic human
traits, such as ignorance, incompetence and stupidity,
to name a few. While these notions are doomed to fail
at actually solving anything, they make up for it by pro-
viding a target for the frustration they create—"them."

Most of us would rather accept the idea of an
enemy among us than to examine ourselves for signs
of the enemy within each of us. In his remarkable study
of fanaticism, *The True Believer*, Eric Hoffer identifies
the desire to join a mass movement as the individual's
desire to escape from himself. "Blind faith is to a considerable extent a substitute for the lost faith in ourselves . . ." he says and later he goes on to warn of the
consequences of this flight from ourselves: "When we
lose our individual independence in the corporateness
of a mass movement, we find a new freedom—freedom
to hate, bully, lie, torture, murder, and betray without
shame or remorse. Herein undoubtedly lies part of the
attractiveness of a mass movement."

Douglas County has changed in the last ten
years. People are much more fear filled now than they
were. Tolerance used to be a hallmark of life here in
the Hundred Valleys of the Umpqua, but now we have
become the land of the anonymous death threat, the

window broken in the night, the job lost because of political beliefs and the whispered accusation that turns neighbor against neighbor—a cowardly land of "us and them."

It is a shameful way of life and one we all bear responsibility for. We have killed our hearts because we found pain there, the pain of hard choices. What we have left now is something inhuman and heartless. We have learned to hate one another. Because we could not trust ourselves to love we have given in to fear.

The price of fearfulness is either the hard struggle to accept uncertainty as the cost of living in an open society or the death of openness in our society—Thomas Jefferson or Adolph Hitler, The Bill of Rights or Kristallnacht. Those are the choices we have to make for ourselves, the choices that will determine our future long after the issues of today have faded away.

1991

## THE SMELL OF HOME

Upriver from Tiller, Ore-
gon, the South Umpqua River pours through a narrow
channel between gray basalt rock faces into a deep, still
pool where salmon circle slowly, waiting for fall rains.
The spring Chinook arrive here in June after a two-
hundred-mile long journey upriver from the ocean.

The salmon know the smell of home, the scent of
jasper, basalt, porphyry, quartz, agate and tufa carried
by the waters from the gravel bars where they hatched.
Patiently, they work their way against the current
returning from the Aleutian Islands home to the South
Umpqua.

They wait out the long summer months—when
the river slows and the water grows warmer—never
eating, living on the fat stored in their huge bodies.
On summer mornings you can see the sore-headed fish
from the cliffs above, silvery ghost shapes in the sun
dappled waters below, moving in a slow, solemn circle
dance.

They are a bruised and battered lot, bearing the marks of their passage, old wounds from seal bites, fish hooks, nets and the scraping of rocks encountered in the riffles of the home stretch. Their flesh, once firm from the arctic feeding grounds, grows soft in the warm river water. Fuzzy white patches appear on their scaly sides, the mark of infection and a sign of approaching death.

They are prisoners here for a while, holding in the deeper pools scattered among the shallow upper reaches of the river, rising in the cool, quiet morning hours and hiding in the depths when the afternoon comes bringing heat and the campers and bathers who splash about on the surface.

Evening comes, and the humans leave. Blacktail deer come down to drink. The firs and cedars cast long shadows across the pool. The clever-handed raccoons fish for crawdads along the edges and silence returns to their watery world with the night.

There is a quiet joyfulness to their languid circling—not the exuberance of their leaping struggle through white water on their way up here—but a deeper joy made of patience, survival and expectation. Their long journey is nearly over, the uncounted thousands of miles behind them. Soon the rains will come and they'll swim upriver on the rising waters as their ancestors have always done, to dig their nests on gravel bars, and lay their eggs in the waters of home.

1991

## DIVING HIGH

The rock is older than the river. Two hundred million years ago it welled up from the floor of a shallow Jurassic sea. Along with a dozen other nearby outcroppings collectively called the Boomer Hill monoliths, it was slowly buried and lost until the Klamath Mountains rose, forming the South Umpqua River. The river cut its way down to the rock and around it, leaving a deep pool at its base. Both the rock and the river are young when reckoned in geologic time. Still, its gray-green face has looked down on those waters for a good long while.

It's a big rock, about twenty feet high and running along the river bank for one hundred feet, all one solid mass. Every year, about the time when the spring Chinook salmon have finished their migration upriver, I take its measure and my own. So far, I've changed more than it has.

There are many diving spots on the river—cliffs and rocks and bridges—all the way from Deer Lick Falls on the headwaters down to the sea. But Boomer Hill Rock is special for me. It's the closest one to my home, the dive I know best. It's also just the right height—low enough for safety but high enough to frighten me.

At middle age, I find that I scare more easily now than I used to. I can remember flinging myself from the rock in drunken, sloppy flips, convinced of my own personal invulnerability. I no longer expect to live forever, and so I approach the dive, as I try to approach everything these days, with more respect.

The first dive of the season belongs to this place. Often it's the first swim in the river for the year—a brief fall into the joys of summer. My feet are always tender after the cold months. The sand is hot and the rocks are sharp. I could swim upstream to the rock from the beach where we lay out our towels and picnic paraphernalia. My son always sprints into the water here. My wife wades in slowly, inching her way into the cold water until it reaches her shoulders and then gliding away purposefully upstream. But, before that first dive, I avoid all contact with the river, not out of superstition but from a sense of propriety, a desire to do this one thing right in spite of all the other things in my life that I botch.

From the beach I walk along the river bank and climb the rock from the backside, cross its top and come to the edge where a stone, about the size and shape of a football caught in the matrix, juts out over the water. It's not all that high of a spot—I've dive from twice that height before. but from up here it looks higher than it really is. The river here is deep, wide and slow, a stretch where it tries to become a lake. The water is green with gold speckles but clear enough that I can see several feet down below the surface, giving the illusion of added height.

There's also an optical phenomenon here, the same one that makes circular rainbows around the shadows of airplanes on the clouds below. Standing on the rock, with the afternoon sun at my back, my shadow on the water has golden rays of light emanating from it. The ancient Hawaiians used to climb to the rim of Haleakala Crater on Maui to see their shadows on the clouds trapped in the caldera below. They called it *akakuanuenue*, "seeing your soul."

I've seen it all many times over the years, the disembodied submerged reef forty feet out in the river, the dappled silver and black light on the surface of the water, the golden halo of my shadow, the half-guessed shapes of fish swimming along the reef below. There's a dream-like state of mind that is brought on by the scene, my own nervousness and a touch of vertigo. It's important for practical as well as spiritual reasons to shake it off before I dive.

Looking up helps. Sometimes there are teenagers sitting on the reef or up on top of the rock; sometimes there are just my wife and son, small figures on the beach downstream, looking up at me. I used to work as a steeplejack, facing the fear of falling every day and often at night in my dreams. There are worse fears though, the kind without names that are known only by the chill they produce, more terrible because of their ambiguity.

There's something very simple and clear and honest about cliff diving that appeals to me even more than the thrill. It's hard to find moments like that in this muddled age.

The dive itself is deceptively simple: Spring out, align the body and fall. I have a friend who teaches kung fu. He told me that breaking bricks with his bare hands feels very relaxing when he does it right. I can believe him, because that's the way a proper dive feels, very natural and effortless. Other than body mechanics, two mistakes can mar a dive—shying away from the water and reaching for it. Both can lead to injury and both are the result of fear. But fear is a projection onto the future, and there's no room here for anything but the task at hand.

I wave to my family, fold my hands in an attitude of prayer and begin focusing my mind on the dive, imagining it step by step ahead of time; the stretch, the crouch, the springing, aligning my body, falling and entering the water—Zen archery, with my body for the arrow. And then I do it.

There is a momentary empty feeling in my stomach, the tingle of leg muscles as I crouch and leap, a weightlessness, the sparkle of light, the roar of water in my ears and a rush of bubbles along my skin as I glide along in the deep, slow down and swim up toward the surface.

1988

## WHO OWNS THE RIVER?

When my wife and I moved to the Umpqua, a big part of the reason we settled in this valley was that there is a river here. We grew up in Los Angeles and having a river was something new and exciting. Swimming, rafting and fishing the river, or just seeing it out our living room window, made us feel like millionaires even though we were barely getting by.

It's easier now to forget just how important the river is to our lives because it is always there, present in just about everything we do. Trying to appreciate it is like trying to appreciate air or sunlight or the roof over our heads or each other.

When our son was seven years old, we flew down to Los Angeles for Christmas. The boy was born in Roseburg, Oregon, and on the flight I was trying to see through his eyes, as fathers do, wondering what he'd think of the overwhelming reality of a major city. Someday, I knew, he'd grow up and decide for himself whether to live in the small town of his birth or to move.

It's an important question in many ways, both for the family and for the place where we live. Four generations of migrations, from Khazakstan to North Dakota to California and to Oregon has allowed our family to survive but left us poorer too, as the miles and the years pile up to cut us off from our past. The valley too suffers when the young leave and the land falls to new people who don't understand its limits. It takes long years of generations living in a place to build both the family and the culture that can allow it to survive.

Sitting next to the boy as we came in to Hollywood/ Burbank Airport, I watched him looking out the window at the San Fernando Valley below. He pointed over toward the Santa Monica Mountains in the west and asked me, "Dad, what's that big gray thing over there?"

"Oh, that's the Ventura Freeway." I told him, "Sure is big ain't it? Look at all those cars. You never see a freeway that big at home."

"No, Dad," he said, "not the freeway—that thing right next to it, the big concrete thing."

"Oh that. Well, that's the Los Angeles River," I told him, and he looked at me in disbelief, then, when he realized that I wasn't kidding him, his face contorted in revulsion.

"That's a river?"

"Well, yeah, sort of," I told him, "At least it used to be, but they paved it back before I was born, back when my big brothers were little. It's just a big storm drain now, dumps all the rain water out to the ocean, but it used to be a real river with fish and trees and all."

The Los Angeles River is a good example of what happens when we see our world as a collection of unrelated things that we can tinker with. What happened to the river also happened to the valley and the people, because, in reality they don't exist as isolated parts but coexist as something much more complex.

The Umpqua River is not a flood channel, or a water supply, or a source of hydraulic power for generating electricity, or a way to get from one place to another, or a playground, or a fish hatchery. It is first and foremost itself.

We hear talk about conflicting interests in water ownership—recreation, wildlife, irrigation, land values and economic development. But it's important to remember that these are all human concerns and that the river itself has its own agenda, one that doesn't recognize any mere human needs.

Who owns the Umpqua? State and federal agencies, urban water and sanitation districts, power companies, farmers, home owners, and sportsmen all lay claim to it. But, really, how can anyone own a river?

Ownership implies control. Have you ever stood by the river bank during the high water and watched the flood roaring past, seen those huge waves carrying logs and debris? Can anybody own the flood? Who owns the sunlight on the water? Who owns the osprey's flight or the ghostly shapes of Spring Chinook salmon circling in deep pools? Who owns the raccoons reaching under rocks for crawdads at night? Who owns the sound of whitewater rapids or the color of maple trees in the fall? They're not yours or mine at all—they're the river's.

In many ways, we've got the whole notion backwards. We don't own the river—the river owns us. We are its people, the people of the Umpqua. Everywhere we go, we are forced to encounter the river. When we give directions we say "upriver" and "downriver" or "across the river" instead of "east" or "west" or "north" or "south"

The river is an inescapable fact of life here. It shapes us even more than we shape it. We need it more than it needs us. It was here before we were and it will still be here long after we're gone. We can dam it, channelize it, pump water from it, build bridges over it, but in the end, whatever we do is only temporary, because the river will do what it's going to do. It has a strength and a patience that are beyond our own and a wisdom that is beyond our understanding.

At best, we can hope to live in some kind of harmony with the river, but that's only possible if we approach it with a great deal of humility. All of the mistakes we've made, the problems we've created, have come from our arrogance in thinking that somehow we understand the river and can manipulate it for our own purposes.

On Sunday morning, February 16th, 1870, the sternwheeler *Swan* arrived in Roseburg. Within minutes church bells were ringing, people were firing guns into the air and a wild celebration broke out all over town. It was hailed as the dawn of a new era of economic prosperity for Douglas County.

The boat had come, inching its way upstream from Scottsburg, at the instigation of the Merchants and Farmers Navigation Company, a local group that had formed to open the Umpqua River to steamboat travel.

At the time, the Oregon and California Railroad had halted construction at Eugene, 70 miles north, for lack of funds. The 1,500 tons of goods imported annually into Douglas County and the 1,000 tons of goods being exported were still being hauled in wagons to and from Scottsburg, a long trip that cost $40 per ton in freight charges. A steamboat could carry the freight for half that price, saving $50,000 every year—if a riverboat could force its way upriver to Roseburg.

Captain Haun declared that the river could be made navigable by blasting a handful of obstructions at a cost of a few hundred dollars. The Navigation Company petitioned Congress for a study and the Army Corps of Engineers sent out a survey party that summer. Despite the fact that the river was impassable even in a row boat by then, the engineers recommended the project. After all, the *Swan* had, impossible as it seemed, made the trip.

Having waved the magic wand of a feasibility study over the project, Congress appropriated $22,500 to make the river navigable—about forty times the amount that Captain Huan had said it would cost. Contractors spent the summer of 1871 blasting ledges and boulders and pulling logs out of the river to carve a boat channel while the Navigation Company built a new riverboat, appropriately christened the *Enterprise*, to make the inaugural run.

When the fall rains came the *Enterprise* set out from Scottsburg. Unfortunately, it never arrived in Roseburg. The new channel had sped up the current and the boat couldn't fight the increased flow. Even if the *Enterprise* had reached town, it was too late by then anyway. In April of 1872, a few months after the new riverboat's unsuccessful maiden voyage, the O&C Railroad line reached Roseburg, making steamboat travel obsolete.

The Merchants and Farmers Navigation Company folded in 1872, three years after it began, but the steamboat channel is still there, and we're still paying for it. For the past 120 years the river has flowed more swiftly, causing greater winter erosion and lower summer water tables. When we look at the consequences of an increased water flow rate—less water storage, more soil erosion, loss of riparian vegetation, siltation, warmer and more polluted water, fewer fish—it's a safe bet that the true cost of the riverboat project has been millions of dollars.

It takes a long time to understand a place, to learn what its limits are. The beaver were trapped out between 1820 and 1840 and with them went the thousands of small dams they built. Hydraulic gold mining silted up hundreds of miles of gravel spawning beds. Moving logs downriver scoured the bottom. Marshes were drained to make pasture. Cattle and sheep grazed off the streamside brush, causing erosion as the banks gave out. Clear-cutting removed thousands of acres of the forest canopy. Following the flood of 1964, the federal government began a program aimed at removing woody debris from the headwaters creeks, turning them into what might as well be concrete storm drains.

All these practices have changed the dynamics of the river. We have always had about the same amount of rainfall, but what has changed is what happens to the water. It gets to the ocean much more quickly now, instead of soaking into the soil and spreading out. Lower summer flows, coupled with less shade from trees and brush, makes for warmer water. Pumping water out for urban water systems and farm irrigation leaves what's left lower and warmer and funkier still. Pollution, from industrial chemicals, sewage spills and agricultural run-offs, becomes even more disastrous when there's less water to dilute it.

It's hard to see the changes happening because the damage hasn't been a matter of dramatic calamities. Instead, it has occurred slowly, bit-by-bit over the past 150 years—a wetland drained here; a grove of trees logged off there; a road cut too close to a stream; a field that has been over-grazed. None of these small events, in and of themselves, is enough to kill a river. But the accumulation of these misuses, over a period of time spanning several generations, amounts to a disaster greater than any flood or wildfire or chemical spill.

If we could see it happening all at once, we'd be much more alarmed. Rust and fire are the same chemical process: oxidation. If your car was in flames you'd call the fire department. But it's easy to ignore rust, even though it'll ruin the car just as surely and completely as a fire, given enough time. It's the same process, the only difference being a matter of pace.

The river's pace is not a human one. Our longest personal measurement, our own life span, is barely enough to begin measuring the rate of change in a river. And, of course, even that requires careful attention, decades of accurate observation.

Our home in Myrtle Creek sits up on a hillside, about 150 feet above the valley floor. From our living room we can look out across the Missouri Bottoms, a small alluvial plain about a half-mile wide and four miles long. The South Umpqua River runs through the valley, and so does Interstate Five, The Southern Pacific Railroad's Siskiyou Line and Old Highway 99. It's the only flat ground for miles around.

The valley used to be called Orchard Valley because it was planted in prunes, back before refrigeration killed off the dried fruit industry. The orchards are pretty much gone now, with just a few remnant patches here and there, but the bottoms are still in farm land, wheat, hay and pasture. It's prime land, loamy top soil anywhere from twelve to thirty feet deep.

The bottom lands look flat but aren't really. When the sun gets low in the late afternoon you can see the shadows cast by a network of gentle swales winding through the Missouri Bottoms, running in the same direction as the river.

It took me a year of living up on the hill above before I noticed them. It's a subtle thing and even when I did notice them I never paid them much attention. I assumed that they were the remains of old river channels, carved as the lifting of the mountains on either side of the valley pushed the river back and forth between them.

One morning, after I'd lived on the place for five years, I went out to the barn to milk our cow and heard a roaring sound. It was dark out, winter time, and it had been raining hard all night. I figured that the noise was from Hilp Creek, running through my place a hundred yards from the barn.

I sat on my stool, milking the cow, and as the sun came up, I began to see out across the valley and realized what was going on. The river had jumped its banks. The high water was running through the bottoms. There was a whole new river running parallel to the main channel and all across the bottoms the swales were full of brown water, smaller rivers intertwining and feeding into the two main flows.

The words "flood plain" came to me and, with them, an explanation for the swales: flood channels cut by the high water. I thought I understood now how the bottoms were formed. But it wasn't until four years later that, almost by accident, I finally learned the truth.

I was interviewing a fisheries biologist for the Forest Service. While telling me about salmon and steelhead habitat improvement projects up Steamboat Creek, he mentioned that the Willamette River, which runs in a single channel today, used to be a braided stream. In the stretch between Eugene and Corvallis it ran, all year round, through a network of nine to eleven major channels, with many side sloughs.

The Willamette Valley was a vast wetland, with beaver ponds, marshes and islands between the main channels. Flooding was an annual event. The river channels were choked with debris which slowed down the river current and spread it out, allowing the soil washing down from the mountains to settle out and build up. The marshes were home to huge flocks of geese and ducks. Thousands of herons and cranes and swans, osprey and eagles lived there. Fish and amphibians and insects provided food.

The Missouri Bottoms below my house must have been a smaller version of the same sort of marsh. For years I'd looked out over that land but it never occurred to me that the river used to run all year round through several shallow channels.

Several things suddenly made sense to me all at once. Old Highway 99 runs along the hillside on the opposite side of the valley. I knew that it followed the old Applegate Trail and that the pioneer trail followed the Indian trade route, but why up on the hillside instead of through the flatter bottoms where the freeway runs? Because the bottom was a swamp—you couldn't walk through it let alone drive a wagon through there.

I understood how thirty feet of top soil had built up. A single channel would never have slowed down enough to leave that silt behind, but a marsh, one that became a seasonal lake every spring, would.

I recalled passages from settlers' diaries which mentioned the huge flocks of waterfowl, so thick that to hunt them they simply startled the birds into flight and fired randomly into the air, bringing down ducks and geese with every shotgun blast. Where had these uncountable thousands of birds lived and why weren't they around anymore?

The South Umpqua, as we see it now, a single channel with banks twenty to thirty feet high, is a modern creation. The marsh was drained for farm land, forcing the water into a single channel which flowed faster and cut its way down through the soil to bedrock.

What's the effect of that? For one thing, the soil eroded from the mountains no longer settles in the valley, it washes downriver to Reedsport where the Army Corps of Engineers dredges it out to keep a navigation channel open.

The wetlands used to store water and stored it much more efficiently than any dam because it was stored in the soil, keeping the ground-water table higher. Along with the storage the marsh provided filtration and cooling, so there's not only less water now but what we have isn't as cool and clean as it used to be.

Of course, we've lost the wildlife habitat too. Wetlands are tremendously productive. In order to make wheat fields and hay ground and pasture we've drained our wetlands, but in terms of sheer protein per acre, the "dismal swamps" were much more productive than the farmland that replaced them.

Some day we may decide to restore the marsh. We might decide that the benefits of erosion control, flood control, water storage and filtration, and wildlife enhancement outweigh the value of the crops that the land produces.

The San Fernando Valley, down in Los Angeles, is also an alluvial plain. Although they've built a city on it, the essential nature of the place hasn't changed. It's still a catch basin for the run-off from the surrounding mountains and every once in a while nature reminds them of that fact with a flood.

Back around the turn of the century, Los Angeles County was the top farming county in California. I can still remember driving through the San Fernando Valley as a child and seeing farmers working their fields while housing projects were springing up all around them. My father once told me about coming to Los Angeles in the 1930's. He used to hunt ducks and geese in the San Fernando Valley marshes. The old nickname for the valley was "The Frog Pond" and people still call it that sometimes.

To see the place now it's hard to imagine how it was when waterfowl wintered there and condors still soared above the hills. But the change came in a single generation, in the thirty years between my father's early manhood and my own.

My father came to a place of beauty to start a family because his birthplace couldn't provide a decent living. I had to do the same thing because my birthplace had become a wasteland of concrete and asphalt and smog and crime. I'm hoping that my son won't have to do it all over again, that maybe this time we've found a place that will stay beautiful and safe.

It's hard to say whether that will happen or not. Dangerous changes come, sometimes unexpectedly and sometimes slowly. But if this place is to survive, the Umpqua River, the living heart of these valleys, needs to be watched over and cared for so it can continue to take care of us, its people.

1992

Nothing seems more permanent here than the mountains, the strong rib cages of the Umpqua Valleys. Rising steeply from the bottom lands along the river they stand fixed with a patience beyond time, a reminder of just how short the days of the men and women below really are. Their forested slopes, the grazing ground of clouds, draw our eyes upward reminding us of things beyond our knowing, of the secrets hidden in their distant blue folds.

And yet the mountains move. They are alive, growing and changing in their own time, so slowly that usually only the most exact instruments can measure their pace.

But sometimes in winter the warm rains come melting the high country snow. Then the mountains move swiftly in the night and we awake to marvel at the suddenness of change.

Culverts spew the running ditches, rivulets turn to streams, creeks become rivers and the gentle river rises brown and huge, transformed overnight into great roaring high water creatures that claw at their banks as the mountain sides slip down to the valleys.

In grains of sand and clay, in pebbles and rocks, boulders and trees, the mountains come down on the flood.

We wake and watch, filled with uncertainty, our own lives no longer fixed and stable when the mountains move. We listen to reports and call our neighbors, trying to find something to cling to in the news of washouts and road closings, weather predictions and warnings. We go out to find our familiar places changed beyond recognition. Finally, we accept the fact that we are caught in something beyond our knowing or control and fear gives way to elation.

The unexpected event forces us to cancel our routines for a while and a carnival air comes with the impromptu holiday. No work. No school. Nothing to do but wait and see what it will bring. All our plans are dwarfed by the giant among us.

We come out to see the whirling waters, to gather along the overrun banks and watch the foaming mats of debris riding swiftly to the sea. Logs shoot through high crested waves, railroad ties and pump houses fly past. Bits and pieces of cast-off trash become oddly significant as they bobble by the small groups of people on the shore. Old timers watch and nod and compare the high water with floods in the past and are listened

to attentively for a change as they assure their audiences that they've seen much, much worse.

1991

# HIGH RISK GROUND

Walk carefully here.

Watch your steps and look ahead. This is tricky ground and a little slip could have serious consequences. It's not far, less than a mile, and downhill all the way. You can't get lost, but you might find yourself in some tough spots you didn't foresee and even the easy parts can be more dangerous than you imagined.

Experience will help, if you have any. Most people don't really know what it's like to walk on a slope that no one would tread unless they had to in order to feed their families. Only love or money could have the power to bring someone here, to risk this hike. Those who have crossed this slope were compelled to do so and for nearly all of them, love had nothing to do with it.

No one would walk here for a thrill, though it's steep enough to give you that tingling feeling at the back of your knees and in your armpits when you look at it, a series of layered shale and sandstone rock faces, ledges and bowls, stacked up to the ridge line like a crumpled brown and gray and green wedding cake. This is the Coast Range, not the Himalayas, and there's no glory in conquering a mountain ridge whose top you can drive to in a pickup truck. There are thousands of slopes like this in Western Oregon and this one is distinguished only because of one brief moment and the sudden tragic deaths of three women and a man.

Get close, if you want to understand. Accept pain as a necessary risk. Distance deceives, leaving the mind prey to the transformative power of words. The words, "mountain" and "tragic" are easy ones to mistake. They seem so simple, so believable and so very trustworthy that we easily forget that our words are just symbols. Everyone knows what they signify, although everyone would, if questioned, admit that the reality of any particular mountain, or tragedy, is much more complex than the words suggest.

The mountain in the mind is no real mountain at all, no more than a contour map is the land itself. The actual mountain is bigger, more compelling and interesting. Seen from a distance, it has a shape and colors but lacks the intimate details you would confront in climbing or descending it, when those far-off silver matchsticks of sun-baked logs become obstacles in your path and the small green splotches of brush swallow you in tangled viny maple thickets, dark and mossy in the shade where sword fern, salal and Oregon grape grow atop duff-covered rocks.

The logged-off slope here above Hubbard Creek mostly runs from 70%–90% grade, or at an angle of 35–45 degrees, with rock faces here and there that are nearly vertical. Believe me, setting chokers on that kind of ground provides plenty of adrenalin rushes.

Working a high-lead logging show on a 70%–90% slope means that you're going to have to jump or duck because something big and mindless is headed your way and moving fast. It could be a log, or part of a log, or a boulder or a stump. You won't have time to think and you won't have good footing. You'll need to jump quickly and jump in the right direction. You rely on pure instinct, quick reflexes, caulk boots and God's will to earn your daily bread.

Logging has always been dangerous work, but today's loggers are working ground that their grandfathers never would have tackled because they didn't have modern yarders. Logging is also faster now, so much so that a typical logger will send more wood to the mill in a twenty-year career than his father and grandfather did in their combined sixty years on the slopes.

How steep is too steep to be logged safely? Paul Bunyon, legend tells us, logged the Upside-down Mountain. Should such a mountain actually exist, there are no occupational safety or forest practices laws here in Oregon that could stop it from being cut.

Search carefully, look everywhere and you'll find that there are no issues lying here among the rocks and soil and brush and debris, unless you bring them yourself, stuffed into the backpack of your mind like rain gear, to be pulled out for protection when things get uncomfortable.

How can you fully know a mountain or a person? Both exist through time, always changing, made up of many sides and aspects and each side, each aspect, a little different from the others. How can you begin to understand a tragedy composed of people and a mountain?

Four people, Rick Moon, Susan Moon, Sharon Marvin and Ann Maxwell were swept away in a debris torrent. The rains came, fell on the mountain and washed away a relatively small piece of it, a headwall divot about thirty feet across. It gained speed and size, sweeping along more water and jarring loose more logging debris and soil, which picked up yet more of the same material. Within a minute or less, traveling a short 3/8 of one mile, it hurled a watery wall of mud, boulders and trees down upon them. The wave carried them, the house, sheds, and a pickup truck along with it. There were four brief moments of human terror and then they too became mere lifeless debris in a churning brown mass.

Something went terribly wrong.

Stand there today and look. At the ridgeline there's a landing, a wide spot in a graveled logging road. At the bottom is a grassy meadow, studded with boulders where scattered muddy bits and pieces of a home lie among wildflowers and red clover. The landing dates back to 1954. Below the landing, and above the wreckage, is some steep ground where a stand of trees took root sometime around 1907 and stood until 1987. The wreckage is of a house that was built in 1972 and sheltered the Moon family until November of 1996.

In the fall of 1986 a forester from the Roseburg office of the Oregon Department of Forestry walked through an eighty-year-old stand of trees uphill from Stump Acres. There was an old landing at the ridge line, some extremely steep ground below the landing, a draw containing the headwall of Rock Creek and, at the bottom, a home sitting on an alluvial fan next to the creek.

Looking around, he noticed that the ground was steep and, judging by the hooked "pistol butts" on the first ten feet of the tree trunks, unstable. He noted the location of the house below and saw that the place was built in a dangerous location. He knew that, should the headwall collapse in a winter storm, part of the mountain would come roaring down the draw and probably destroy the house. He also knew, full-well, that logging on unstable ground increases the likelihood that a landslide will happen.

Depending on how competent he was, he may or may not have been aware of just how far behind the science of forestry the forest management laws and rules he was working with actually were. The groundwork for what is acceptable practice today had been laid-out in 1979 and fleshed out over the next five years by a group of researchers which included Jerry Franklin, the "father of the Northwest Forest Plan."

Franklin and another member of the team, Chris Maser, toured the Northwest and spoke at national and international forestry conferences. Their message, that current practices were destructive, caught on immediately in Europe, Asia, Central America and South America. But close to home, in the Pacific Northwest, their words met with stiff resistance. When Chris Maser addressed a gathering of the Society of American Foresters local chapter in Roseburg in 1985, he was struck by what he later likened to the early stages of people responding to death—with denial, anger, bargaining and depression. In May of 1986 Maser estimated that the science of forestry was ahead of the field practice by ten years.

The scent of change was in the wind. Some foresters welcomed the notion of a "kinder, gentler" style of management, with a more holistic approach that included leaving buffer strips along streams, smaller clearcuts, less slash burning, and leaving snags and large and old rotting logs in place. Most of Oregon's foresters, particularly those working for large corporations, ignored the advent of these "new perspectives" or dismissed the messengers as mere tree huggers masking their political agenda with questionable science.

In 1986 the political and legal battle over forest management was looming large in timber industry and environmental circles just as timber harvest levels in Douglas County were reaching record highs.

The state's forester met with Rick Moon during the planning stages to discuss Rick's concerns about preserving his water supply. His notes don't show whether he ever told Mr. Moon that his house was in the path of a potential debris flow. According to Rick's neighbor, Gordon Marvin, neither he nor Rick, were told about the possibility. Considering the agency's culture in those days and the emotionally-charged politics of logging it's not hard to imagine the forester keeping that awkward bit of information to himself.

More difficult to imagine at such a moment is the chance that they, worried residents and the government's agent, would meet without distrust, as troubled neighbors revealing what lay in their hearts, all three speaking frankly about their fears and hopes, for themselves, for the land, for the future. Difficult to imagine—the possibility of love.

The forester had no legal power to declare any part of the land off limits to logging, and he had no legal obligation to warn the Moons that they'd chosen a risky site for their home. The forester duly noted Rick's concerns and his own belief that the house stood in danger and then wrote out a prescription for logging which was based on existing rules that were already outdated and soon to become obsolete.

By the end of 1987 the Oregon Forest Practices Act had been strengthened, Roseburg area environmentalists were receiving death threats, and the land above Rick and Susan Moon's house had been logged off.

━━✦━━

Most people will, if asked, assure you that when it comes to their job, they are conscientious workers. "I show up on time every day. I'm efficient and reliable. I follow the policies, pull my own weight, give an honest day's work for my pay," they'll tell you. Yet, nearly everyone will admit to having occasional moral qualms about their work.

"A corporation has neither a backside to be kicked nor a soul to be damned," as people have been saying for so long that no one remembers who minted the adage. Henry David Thoreau probably knew the saying back in the 1840's. In "Civil Disobedience" he wrote, "It is truly enough said that a corporation has no conscience; but a corporation of conscientious men is a corporation with a conscience."

Most people will assure you that they are conscientious citizens. "Sure, I obey the laws," they'll say, "I vote. I try to help out where I can, be a good neighbor and all that kind of stuff." Yet, if you attend any town meeting or public hearing where thorny issues are at stake, you'll see that few citizens show up and fewer still speak up. And from those who do speak, you'll hear mostly half-truths and evasions, jargon and ideology, bitterness and rancor.

We have sought to protect our people and our land through laws. Yet, in this case, everything that was done was utterly legal. Laws come about as a reaction to failures of conscience. We give free reign to people, businesses, institutions and agencies until their actions cause harm. It is only after the fact, when it becomes clear that someone failed to apply the universal moral standard known as "The Golden Rule" that legislation becomes necessary. "There ought to be a law," we say, and eventually, if the abuse is obvious and common and talked about enough, our legislature responds.

It has been obvious for decades that not every place that can be logged should be logged. Industrial silviculture, or "tree farming" as it's been called, is like Longfellow's little girl with her little curl: when it's good it's very, very good—but when it's bad, it's horrid.

The tragedy was swift and simple. The headwall of Rock Creek lost its integrity during an unusually heavy winter storm. The same thing happened that November day on dozens of other steep slopes in the Coast Range of northern Douglas County. The same storm loosed a landslide that swept a woman and her car off state highway 138 and into the Umpqua River.

Within the next sixty days two more high-water storms hit the county and each time dozens of unstable slopes came down on the floods. Other homes were damaged by mudslides, roadbeds slipped, culverts washed out, and county and state road crews worked overtime repairing the damage.

———

"Clearcuts Kill!" the protestors' picket signs read.

"They had it coming," a local logger told me.

Between these two partial truths lies a tangled emotional and intellectual thicket of grief, fear, anger, denial and technical jargon. We have met tragedy and responded with lawsuits, studies, hearings, protests and editorials. Four people died and, so far, no one has apologized.

It's doubtful that anyone ever will apologize. Somehow, while the numbers were being crunched, the experts questioned, the briefs filed, the depositions taken, the ramifications of potential legislation considered, and the economic and environmental aspects of logging debated, whether or not human decency has anything to do with public safety was never considered.

1998

## HONKERS

The morning fog seems to magnify sound. Roosters crow at the pale disk of the sun and sound too near, as if the hen house had somehow moved closer in the night. The familiar view becomes mysterious, the half-seen shapes of trees down by the river float like dark disembodied autumn spirits.

On fall mornings it's easy to believe in the dream world, that half-guessed realm which lies concealed inside the everyday landscape. Are those brown-cloaked druids filing by in procession or just the neighbor's cows heading out to pasture? What is it about morning fog that turns a telephone pole, mute and alone, its wires connected to mist, into a looming symbol? Which is the real world, this uncertain morning or the day made familiar by sunlight and coffee?

Geese fly by, Canada honkers, in pairs or gaggles, noisily laughing in their awkward flight, joy filled nomads heading for breakfast in the wheat-stubble fields by the river.

It's hard to stay somber after the honkers fly by. You can't help smiling listening to their mocking laughter, watching their flailing flight, necks outstretched like sprinters nearing the finish line tape.

They take their fun seriously, these heavy bottomed feathered gypsies passing through the valley. "Behold the fowls of the air; for they sow not, neither do they reap, nor gather into barns . . ." and they manage to have a good time of it too.

You seldom see a solitary goose or gander. They seem to know that real fun is sociable. What good is a joke that isn't shared, after all? And they seem to find plenty to laugh at, winging by overhead, trading one-liners like a nightclub filled with comedians, while we stand below, outsiders grinning upward, wondering if the joke just might be on us.

Maybe they do laugh at us earth-bound two-leggeds weighed down with our mortgages and our search for meaning, leading our "lives of quiet desperation" on our lifelong journey from the obstetrics ward to the intensive care unit. The panorama of the human comedy, spread out below, would be enough to amuse generations of those with wings.

Or maybe they just laugh for sheer joy at being geese, delighting in the world spread out like a vast picnic before them, a noisy boisterous tour group passing through, stopping off to enjoy the season before taking wing in great v-shaped flights to other places.

1991

# ANIMALS

The boy was pumped this morning on the ride to school. My son has his first game today— the Coffenberry Junior High Hornets vs. John C. Fremont Junior High Generals. He's on the Coffenberry seventh grade football team, playing first string offensive tackle, second string linebacker and blocking on the kick-return squad. The coaches told the boys to wear their jerseys to school today so the kids could "see who the studs are".

He was sitting there, wearing his black jersey with the gold #79, muttering "We're gonna kill 'em." with this wild look in his eye. I could tell he was worried about the game too, going up against a big city team. "We gotta beat 'em."

To my amazement he weighed in at 4'11" and 122 pounds at his physical—I didn't weigh that much until I was 18 or 19. But he has lost some gut in the last three weeks, down to 118.

"Jeeze dad, they ran us 'til we almost puked."

"Yeah, how 'bout that? Who would've imagined?" I'd been telling him all summer to get in shape, repeating ominously, "Hell Week's coming son—you'd better be ready." But I don't think he really believed me.

It never occurred to me that I'd be the father of a football player. My brothers all grew to be short and stout, like my father and grandfather, but I was the runt of the litter, short and skinny. As a 104 pound fourteen-year-old who was better at writing sonnets than sacking quarterbacks, I suffered through a season of Physical Education touch football with the nickname of "Animal" as in, "OK, so that leaves us with Animal, so you guys have to kick off and we receive."

Coach Fred was the head football coach at my high school and I still shudder at the memory of having him for our PE teacher. "There's only two kinds of kids here," he told us on the first day of our freshman year. "There's the Studs and the Duds. The Studs play football and they don't have to work out in this class. The rest of you are all Duds and you're going to wish you were dead."

He worked us unmercifully and as the school's premiere football season went on, piling up loss after loss, we duds came in for increasingly harsher treatment, push-ups and sit-ups in the muddiest part of the field and endless miles of running.

His favorite target was a friend of mine named Jan Mankowski, a burly kid with biceps bigger than my thighs. I made the mistake of lining up opposite Jan one day during a PE touch football game. He knocked me a full five yards, from one chalk line to the next, airborne every inch of the way. Coach Fred had spotted him in the hallway on the first day of school and immediately tried to recruit him. "Forget it," Jan told him, "football's stupid. I'm going out for shot put."

One of my older brothers played four years of high school football, breaking each of his ankles twice. Years later, in a beery middle-aged conversation, I asked him about those days, whether it was worth it to go through all those injuries.

"Oh, it was great!" he said, "I couldn't believe it. I felt like, 'Wow, you mean I get to run around and smash into people and not get in trouble for it? All right!' The coaches loved me. I was in heaven."

Last spring, I was coaching first base for my son's Little League team and Kurt got tagged out coming in to home plate. I could see him muttering darkly in the dug out so when he took the field for the bottom of the inning I called him aside. I explained to him that it was the manager's mistake, that he should have been held up at third, but he took the routine play as a personal affront and was mad at the catcher for tagging him.

"I'm gonna get him," he was muttering.

Two innings later, he came in to the plate low and sideways, a perfect body block, knocked the catcher for a loop and scored a run. The umpire should have called him out, since Little League rules wisely require runners to slide in that situation, but the ump blew the call and Kurt strutted to the dugout amid parental cheers and high fives from his team mates.

I called him over to my coach's box and, struggling to keep a straight face, chewed him out. "That was a Major League play son, but this is Little League and you can't do that—even in high school ball you have to slide. You should have been called out."

"Yeah, I know," he said, "but I just wanted to cream him so bad."

"Well, don't ever do it again. Somebody could have been hurt."

"OK," he reluctantly agreed, "but it sure felt good."

Later, on the ride home from the game, he confessed, "You know Dad, I like baseball and all but I think football's really my game."

———

I picked him up at football practice the other day, the first time that I saw him suited up. Most of the kids look like kids, skinny arms and narrow shoulders with pads that stick out too far, but Kurt looks like an honest-to-God football player scaled down.

He was walking down from the field with two of his buddies, all sweaty and tired from practicing in the

100-degree heat, the last ones in. A lanky hillbilly, about 18 or 19 years of age, walked by and told them to hustle.

"Why don't *you* hustle?" one of them asked as he walked off.

"Who me?" he shot back.

"We've been hustling."

"Yeah," added one of the kids, after he'd gotten a good forty feet farther up the hill, "Why don't you come down here where I can hit you?"

"What'd you say?" The young man turned around and started stomping toward them. I guess they must have felt pretty big in their pads but they suddenly looked pretty small with this gaunt snaggly-toothed six-footer coming at them.

They tried to walk off but he kept bearing down on them so I yelled over to Kurt, "Hey Kurt-o!" just to let the guy know I was there watching.

"Yeah Dad?"

"I'm picking you up today. Meet me up here. OK?"

"OK."

I didn't get out of the truck, just sat there watching and being ready and hoping he wasn't a speed freak or some other kind of maniac. But it was one of those "What'd you say?"—"Oh yeah?"—"You watch your mouth, kid," kind of confrontations, like house cats hissing and puffing up until somebody backs down and slinks away.

The boys slouched off to the locker room and the hillbilly stopped by the truck on his way back up the

hill. His face was thin, with acne scars and Elvis side-
burns. He looked like one of those perverts whose mug
shots you see in the newspapers.

"Punk kids," he offered by way of explanation,
"Somebody's gotta teach 'em some respect," which was
true.

"I reckon," I offered, wondering how much self-
respect a guy who feels threatened by twelve-year-old
kids could have.

He lit a cigarette. "You want one?"

"No."

"Fuckin' punks," he repeated and walked off scowling.

I sat there in the truck thinking about growing up
and testosterone and puberty and the mating battles
of bull elk and remembering my father, surrounded by
his sons, watching television through a cloud of cigar
smoke as the Los Angeles Rams butted heads with the
Greenbay Packers.

I thought about how all-important learning to be
a hard-ass once seemed and how glad I was that those
days are long past for me. It's probably a good thing the
boy doesn't know what's ahead for him, I decided.

Kurt came up from the showers and got in the truck.

"How'd practice go?" I asked.

"Great!" he said, "We scrimmaged against the
eighth graders. I thought they'd be tough but it was
easy. We killed 'em. They're a bunch of woosses. I didn't
let one guy get by me."

1992

## THE FIELD OF REALITY

The sports writer for the big city paper sat up there in the press box all weekend long mouthing it: "ROSE-burg"—like that, with 90-weight contempt in his voice and a sneer on his face. It got on my nerves real quick.

It got to where I wanted to cold-cock him with a whiskey bottle but all I had was a can of pop—and that just wouldn't have been right. I mean, if you're going to do something like that, you need the proper tool.

I ought to be used to it by now. I've lived in this place for just shy of twenty years now, and I've been hearing that tone of voice whenever I leave the county, for all that time. I know what they're thinking when they say it like that: ROSE-burg RED-neck.

Yeah, well, OK, redneck country, I guess, for lack of anything better to call it. I won't deny that. But, here we were: sitting in one of the finest amateur ballparks in America, watching the 1993 American Legion World Series unfold below us, and it never occurred to him to wonder why 120 of the most gifted young baseball players in the country were competing for a national championship in a little backwater town like Roseburg.

❦

Three cities currently host the American Legion World Series, all of them small and not noted for much of anything else: Roseburg, Oregon, Fargo, North Dakota and Boyertown, Pennsylvania. Roseburg, naturally, I know well. As for Fargo, I've never been there, but my folks come from North Dakota, so I can imagine it easily, the type of people who live there and what they do to earn a living and pass the time.

Boyertown is a mystery to me, though it must be small because it's hard to find a map with Boyertown on it. In some ways, at least, it's probably like the other two towns—small, nothing much to do, a place where most of the inhabitants know each other by sight if not by name.

So why has the American Legion picked these little places to host their national championship tournament? Is the Legion itself essentially redneck?

❦

I must admit that the phrase "100 percent Americanism" that adorns their statement of purpose puzzles me. How do you quantify being an American? By what standard of measure? Is it possible to be less than 100 percent Americanistic—say, 92 percent or 73.5 percent—and therefore unacceptable?

The phrase brings to mind the sort of mentality that, during the hysteria of World War I, forbade my sod-busting ancestors from speaking their *muddersproch* in public assembly, both here on the Great Plains of America and on the steppes of the Ukraine (where the goal was probably "100 percent Russianism").

But it has been a long time since the days when Legionaires lurked in dark alleyways outside union halls, axe handles in hand, waiting to pounce on Wobbly agitators. None of the smiling old men at the tournament seemed capable of anything more violent than a bit of mild ribbing directed at one of their gray-haired buddies. The phrase, "100 percent Americanism" seems to be an awkward anachronism, a leftover slogan from a less-sophisticated age, and I never heard anyone actually utter it during the long weekend.

So, the question remains, why hold their biggest tournament, the national championship for a system of some 4,400 amateur baseball teams in a little place like Roseburg, Oregon?

—⟨✦⟩—

Glancing through the record book in my press packet, I found a clue. Between 1925 and 1954 overall attendance at the American Legion World Series tournaments topped 40,000 fans six times and never again since.

Since 1955 it has reached over 30,000 ticket sales six times: 1974 and 1993 in Roseburg, Oregon, 1976 and 1977 in Manchester New Hampshire, 1982 in Boyertown, Pennsylvania, and 1990 in Corvallis, Oregon. The 1993 series attendance reached 34,306, the highest of the six, and had Medford, Oregon, won the final game of the regional tournament in Billings, Montana, instead of losing by a score of 4–2, it might have topped 40,000.

Roseburg, Manchester, Boyertown, Corvallis— not exactly what you might call teeming metropolises.

More telling though, was the attendance records from the two series held in New Orleans: 8,438 fans at a single game in 1937, tenth highest on the list; 374 at a game in 1984, the lowest turn-out ever.

The 1937 series in New Orleans drew a total of 22,726 fans but when it returned there in 1984 only 7,765 bothered come out to watch the games—less patrons for the whole series than had come to a single game forty-seven years before. Somewhere between 1937 and 1984, the city of New Orleans, Louisiana (along with the rest of urban America) let baseball die.

---

The decline of Major League Baseball (and you will find no one who knows and loves the game who will tell you that it hasn't decayed considerably) was preceded by the decline of amateur baseball. Oh sure, the numbers look good, more kids than ever playing in more leagues, more money spent on more equipment, more and better fields these days, and yet, a whole generation of kids is growing up who've never played unorganized sand-lot ball. It's not a matter of numbers and dollars, but of something no Certified Public Accountant could show on a spread sheet—love of the game.

My colleague didn't know it but, down below us there, watching every game, every day, was old Frosty Loghry. Each day, Frosty sat surrounded by a different set of middle-aged men, all of whom played for him at one time or another. In forty plus years he has raised two entire generations of young ballplayers who went on to become coaches and umpires, league presidents and commissioners and, of course, the fathers them-selves of young pitchers and catchers and fielders, kids who play for the same teams on the same fields as their fathers and grandfathers before them.

I don't think that happens in Portland or Seattle much. If Portland had a sense of tradition like that, would they let the Beavers, the last of the old Pacific Coast League teams, move to Salt Lake City? And Seattle? Well, the less said about the Mariners' troubles drawing a crowd the better I guess.

Every day I stopped by and shook Frosty's hand on my way up the stairs to the press box. His guests over the weekend were a *Who's Who?* list of Douglas County baseball.

"Well, Frosty, I see you been keeping your bench plenty full," I kidded him one night.

"Well," he confided, "they keep me warm."

⚬

Of course, the high-toned sportswriter didn't know who Frosty was or anybody else around here. He didn't recognize Bill Gray, the tournament director, either, because he looked right past him like he wasn't there while Bill was emptying out the waste paper basket in the press box Sunday afternoon.

"Hell, Bill, they got you doing everything around here, don't they?" I kidded him.

"Yeah, well, I saw it was full," he explained, a little embarrassed to be caught in the act, and then he grinned and winked at me, "Besides, we want to keep you media types happy, you know. A little good press never hurts, after all."

Seventy-two years old, in poor health, literally risking his life to bring the series to town and run the tournament, and he was up there emptying the trash to keep us comfortable. He was enjoying himself too, you could tell.

I'd done an interview with Bill Gray a few weeks before. He didn't talk much about his health, except to admit that it wasn't as good anymore as it used to be. He talked instead of the sixty-eight year history of Umpqua Post No. 16 baseball and his own thirty-three years as a commissioner.

Sitting there in the shaded stands behind home plate, looking out across the sunlit field, he talked about the work that had gone into the facility over the years, about the truckers who hauled in crushed pumice for the warning track all the way from Windigo Pass, ninety miles upriver, and were paid in pizzas and beer.

He talked about the new restrooms that volunteers had put in at cost of one quarter of the contractor's bid. He reminisced about getting 600 old seats from Balboa Stadium, home of the Pacific Coast League Padres in San Diego, when the park was torn down. He talked of lumber, paint, lights, and groundkeeper's carts, and how, through all the years, he'd never been turned down whenever he asked for something.

"But that's kind of how this whole thing has come together," he said, "It's been a work of love by a lot of people, and a lot of interest. It's a good baseball community. We sit down here by ourselves and we support that which we have."

The fact is that, despite all our many peculiarly Umpquan faults, this is great baseball country. For one thing, redneck kids, the sons of miners, ranchers, farmers, loggers and lumber mill workers, make good ball players. For another, it takes a special kind of place full of the right kind of people to keep the game living generation after generation.

American Legion Baseball in Douglas County is as old as the national program itself. Umpqua Post No. 16 has kept it in continuous operation since 1925. The post runs three single A teams and one AA team—the Dr. Stewarts, known as The Doc's. It costs about $80,000 per season to maintain all four teams and none of the American Legion posts in any of America's largest cities supports a program of that size.

Season tickets for the 600 reserved seats that came here from San Diego are sold out every year. To get one you have to wait for someone to die.

According to the program book which Helen Lesh (season ticket holder since 1953) put together, the Doc's won 1469 games and lost 524 between 1954 and 1991, a .737 win-loss average over thirty-eight seasons, including state and national playoffs, regional tournaments and world series play.

Everywhere I went that weekend, I was talking to local baseball people, good old boys and girls, working hard with rakes and mops and toilet plungers and having a good old time. It would be easy to go on with examples of how hard everyone worked to bring the series here and to stage it, to go on and on about what a remarkable place this is. But falling into to some schmaltzy "Field of Dreams" hyperbole would be just plain wrong. What I found was better than any dream, it was a field of reality, with all the unexpected beauty and the grittiness of everyday life.

Maybe baseball itself is inherently redneck, or, at least, something that requires an old fashioned, unsophisticated way of living that began dying out in most of America during the post-war boom years of the Eisenhower administration and lingers on only in backwater towns.

The thirty years following the Second World War brought a shift in American demographics. We are no longer a rural society with a large blue collar workforce in the cities. Instead we've largely become a nation of white collar urban and suburban people.

Of course, many of the old time values are best left to oblivion. We can't afford racism, sexism, authoritarianism and nationalism anymore (not that we ever could). But we still haven't managed to replace those social controls with anything more useful and humane and so the future remains in doubt.

Will the Information Age produce good ball players? Will the Nintendo generation forsake the honest dirt and grass fields for the video screen simulacrum? Who can say? "It ain't over 'til it's over," as one of the game's wisest men once said. But maybe baseball, like the family farm, is part of a nineteenth century way of living and, as such, doomed to continue fading away as we enter the twenty-first century.

---

That the Umpqua is remarkably good baseball country is really a very sad thing—for the game and for the nation. For nearly one hundred years, every town in America had a local team which was as important to the town as the church or the saloon or the bank. These amateur and semi-pro teams kept alive local community and family baseball traditions. Many Major League players' baseball roots come from one or more of those teams. My cousin, Mark Holzemer, who pitched his first game in the majors that weekend, is one of them.

During the thirties and forties and up until the early fifties, Amidon, North Dakota, the county seat of Slope County, had a population of about 120 people. The town had a team of farm boys and young men, the Amidon Rangers, who played against neighboring towns, upholding the town's reputation for good baseball. My Uncle Bud Holzemer played third base, his brother Red played first, my father filled in occasionally in the outfield.

These young men, who played amid the wheat fields, went on to become the fathers of dozens of young ball players and, eventually, the grandparents of more dozens of players.

The Amidon Rangers played their last game in 1953, at about the same time when American Legion World Series attendance began dropping off. People moved off, unable to scratch out a living and the town team's demise came early on in that decline. Today, Amidon's population numbers 29 souls.

We have a long tradition of talking about the curative powers of baseball. Writers and baseball officials often speak of it in the sort of mystical terms which medieval alchemists used in portraying the Panacea, the universal remedy which would cure all mortal afflictions. The American Legion's National Americanism Commission falls squarely within this tradition. The program will, we are told, "combat juvenile delinquency," "build our nation's future" and help develop "a feeling of citizenship, sportsmanship, loyalty and team spirit."

But the game really doesn't need justification at all. Ennobling it with lofty claims of social benefit is really a disservice. The primary reason the game caught on and has survived is that it's a hell of a lot of fun to watch and play baseball—though you'd never suspect that from a reading of the program's purpose and goals.

Such talk misses the point in another way. Though the game can certainly help the community, it cannot survive without the community's support. Baseball can only pay back what's given to it.

Ecologists talk of indicator species, creatures like the northern spotted owl, whose population's rise and fall are a measure of the overall health of the ecosystems without which they cannot survive. Baseball, and especially amateur baseball, its purest form, is an indicator of a community's health.

The number of communities which support a strong amateur baseball programs, not just with dollars but by passing along knowledge and above all else, love for the game, is a measure of the nation's health. When baseball dies out in a community it means that there is no community there anymore, just a place with a name, some fixed boundaries and a rootless, fragmented population.

1993

# NINOS

Ashley had something she wanted to tell me. All the other kids were hurrying out the classroom door for free play but she came over and looked up at me with her four-year-old "I've-got-a-secret" eyes.

"What's up honey?"

She beckoned with her finger and I lowered my ear to her. "I love you," she whispered. Then she looked me in the eyes, smiled and scurried out the door.

It was spring time outside. Sunlight and the laughter of children at play filtered through the venetian blinds and poured in through the open door. I straightened up and stood for a moment, caught in the present, overcome by joy and humility, and wondrously alive again.

I tell stories to kids (and big folks too—when they'll listen). Witches and wise women, wicked wolves and wily coyotes, dragons, treasures, and the Land of Faerie are my stock in trade but I was overwhelmed by a bit of real magic more potent than anything in my tales.

It wasn't the first time that a little one had done that. I work with two local Head Start classes of four and five year olds. The kids are mostly from poor families. Many of them have no men in their lives, a need that almost smothers me at times as they cling to my hands and crowd into my lap during my weekly visits.

At times it's painful for me to know these children and care about them, knowing that, in the end, they're someone else's kids. Some of them carry an unconscious weight about them, the heaviness of defeat learned before their ABC's. A friend of mine, who says he sees auras, describes it as a cloudiness obscuring their luminescence. I don't have the knack of seeing souls, but there's something about the notion of children as small orbiting suns that appeals to me and something painful in the thought of their glow being eclipsed by the larger, denser planets of us big folks.

Every year, when Halloween is approaching, I tell a Shasta story about a man who journeys to the Land of the Dead to fetch his wife back to the village. He lacks the strength to carry her home and fails in his quest, dies himself, and is reunited with her in death. It is a sad story, of course, as any good story about grief must be.

One year, after telling it to a classroom of second graders, Misty came up to me and told me her own story: "The night when my grandpa died, mama drank a whole bottle of schnapps and she passed-out. My dad had to put her to bed. In the morning, the ambulance men came to take Grandpa away and I saw them carrying him out."

She told it so simply and quietly that I was a little stunned. For a moment, I just sat there seeing the knowledge of grief in her eyes and searching for a reply. I could picture it all, the old man a sudden corpse in the bedroom, her mother's tears, her father tenderly carrying her mother off to sodden slumber, the strangers with their gurney arriving in the morning, and the little girl watching it all in wonderment. What can you say to a child who has confronted the mystery of mortality like that?

It's tough sometimes, not knowing, really, what to say to a child. They expect wisdom that we grown-ups don't always have. The stories contain great wisdom but the teller is generally just as foolish as everyone else.

Children seldom see the "quiet desperation" that underlies adult life. Of course, we try to conceal it from them out of mixed desires to protect them from uncertainty and to protect ourselves from insecurity, because, despite our posturing, we're just as confused about life as they are.

I said the conventional things. I told her about my own father's death and that death is natural, just something we all go through and that people don't really die because we remember them.

"Mama says Grandpa's in Heaven with the angels and he's happy," she told me. And though I don't really believe in pearly gates and streets paved with gold or everlasting hell-fire and brimstone, I told her that her mother was right.

Looking back on it, I'm not sure whether she needed reassurance or just a chance to talk about death, something we don't talk about as a rule. Either way, she seemed relieved—almost as relieved as I was.

Several years ago, when I first began telling stories, one class had a mildly retarded five-year-old boy named Arlen, who glommed on to me right away. After a few weeks, I met his mother who had the same distracted, fog-bound look in her eyes. I wrote his behavior off to genetics and the odd way that kids, dogs, drunks, women in trouble and crazy people always seem to be attracted to me. But talking to his teacher I learned that his father had flung him against a wall during an infant crying fit, permanently damaging his brain.

One day, he pulled me aside and whispered, "I'm the only one who loves you Bob." It was an eerie moment. He said it with such sincerity that I half-suspected he was right. I laughed, a little embarrassed, and thanked him, thinking that he was trying to get some attention by conning me. I assumed that he meant, "These other kids are just jiving you but I'm the only one here who really loves you, so ignore them and pay attention only to me." Later though, I read an interview with Alice Miller, a Swiss psychologist who studies child abuse, and I wondered if he was really saying, "You're the only one who loves me."

1993

## CENTRAL HEAT

We can count ourselves lucky that so many ways of doing things that are obsolete elsewhere linger on here. We may be part of the twentieth century but we understand the older pace too, the slower rhythms of a life that moves to the long cycles of nature. Umpqua kids understand the old characters in fairy tales; goats and foxes and ravens don't need to be explained. Old clichés such as "making hay while the sun shines" or "a tough row to hoe" are still self-evident and very much alive for us.

One of those old-time phrases that still makes sense locally though it has died out for most Americans is "hearth and home." I've heard city folks pronounce it "heart and home" and though they may not know what a hearth is, they're really not too far from the truth.

From the first morning cup of coffee to the final light-switch tour at night, we keep checking on the wood-burning stove. We listen to the roar of newspaper, junk mail and kindling waiting for the sound that says it's time for cordwood. We adjust the damper and add pieces at the right time to keep the fire just right— hot enough to burn clean, slow enough to last. Like the family itself, it takes care and patience and skill to provide warmth without too much heat.

With every trip to the firewood pile we take stock of our stored warmth, counting the days until spring piece by piece, secure and complacent or worried according to the height of stacked rows of madrone, oak, maple, fir and cedar. The pieces sound like baseball bats as we load them in our arms and when we return the faint smell of smoke greets us at the door.

The old wood-burner sits in the living room and draws us irresistibly to it on winter mornings. Even the cat can't resist and lies sleeping nearby all day. As we go about our busy day we keep returning to the same spot, over and over again like comets returning to the sun after traveling the long cold void of space.

We stand and stretch out our hands toward the warm iron and steel and turn slowly from front to back, absorbing warmth and comfort and, perhaps, a bit of courage too. It's a cold world out there, after all, and nothing helps chase away the damp like the glowing heart of home.

1991

## EDGAR AND THE CLUSTER

It was December and a time of bad news and I was waiting for my son's school bus. The bus won't come to the house any more. Our little gravel road is too narrow and winding to be safe, they say, and there's the school budget cuts, of course, so he gets dropped off a mile north at old George Roy's driveway at a quarter after three.

When he was five his main ambition in life (except for wearing a cape and leaping tall buildings in a single bound) was to ride the school bus like the big kids. Now, at thirteen, he gets suspended from the bus for rowdiness once or twice a year and wants to be seven feet tall so he can slam-dunk a basketball.

I was early and sat in Edgar waiting for the bus. Edgar Alan Datsun needs a new pair of front tires, a tune-up, a fuel filter and a new clutch. But it's a hard time of year to come up with extra money. Oh well, the small get by; to keep under is to endure.

The old flap-fender pick-up will undoubtedly keep running. In the eight years and uncountable (no odometer) miles since I inherited him, Edgar the Haunted Truck has never quite given up the ghost.

He's a benevolent ghost, named Steve, and I'd hate to lose him since he's been a great comfort to me over the years, though at first I was a little leery of him. For four years, before Steve Hoeffler killed himself breathing exhaust fumes in Edgar's cab, he and I were friends. So when his brother and sister showed up from California to dispose of his bright red corpse they gave me his pickup.

It was awkward at first, accepting such a grim legacy, and I almost refused. But we were broke at the time and couldn't afford to be too proud. White Man's Burden, my 1961 GMC, only got eight miles to the gallon with its monster V-6 engine, so getting one of those little Japanese rigs for free and selling off the Gemmy was just too good to pass up.

Edgar was running rough when I got him and when I discovered that one of the spark plugs was defunct I laughed and shook my head, saying out loud, "Steve, you dumb-ass motherfucker, you were only running on three cylinders." It was the first time I'd spoken to him since he'd died and I felt better.

I dreamed I saw Steve Hoeffler one night—alive as you or me. He was sitting in my living room, quiet as usual, but clearly enjoying himself. I knew he wasn't supposed to be there so I tried to grab his hand but there was nothing there.

"But, you're dead!" I said.

"Yeah . . ." he admitted a little sheepishly, "but it's no big deal." He smiled and then I knew that he was truly dead but he was still my friend and it really wasn't anything to worry about. I woke up feeling better and later, when I went out to the truck, I noticed that the little red EGR light on the dash was always lit because the Exhaust Gas Recycling system was disconnected, so I named the truck, patting him affectionately on the dash and saying, "OK Edgar, let's go."

It's become a habit now over the years, patting the dash from time to time to urge the old beater on or to thank him for pulling through for me despite my chronic neglect.

---

I thought about Steve and his permanent solution to his temporary problem while I waited for the bus. Maybe the dead do give us gifts, like the old tales say. I know I needed one that afternoon even if it was just a matter of having something to compare to my immediate problems.

There was the cluster for instance, which was scary enough, and now the boy was in trouble at school, suspended for three days for theft. The vice principal had called me just an hour before to tell me that my son and a buddy of his had stolen money from a teacher's purse. He's a bright kid, good natured and likeable, but he's thirteen.

So far this year five teenage boys have killed themselves in our end of the county, three by gunshot and two by hanging. A sixth boy, thirteen years old and one of my son's buddies, shot himself in the right temple with an antique revolver two weeks ago but will survive with the aid of a glass eye, reconstructive surgery and counseling. Add him and the other boys to the eleven girls who have also made attempts and you've got a real problem.

It's called a teenage suicide cluster and it's an epidemic that can't be diagnosed until it's too late to do much about it. Everyone wonders "Who's next?" and everyone has a theory: mass hysteria, deteriorating family values, the spotted owl, drugs, domestic violence, too many poor people, Satanism, rock and roll, secular humanism.

Four long distance phone calls gets you in touch with a New York City researcher, one of the nation's leading experts. She gives you a straight answer based on years of study funded by government grants, "We just don't know. We have some theories but no one really knows why these things happen."

Tuesday afternoon, waiting for the bus, Christmas vacation starts Saturday and I wonder, "Who's next?"

"Shit Bob," my buddy Jim, over in Prineville, sighed over the phone, "mass ennui—just giving up. Wanting to get laid and being afraid of it too, and masturbating, the guilt and the doubt and the hypocrisy of school and all the shit you go through—that horrible feeling, sitting in Limbo, loneliness and restlessness and 'What's the use?' . . . Hell, I remember, there were times I thought about snuffing myself, lots of times. You did too, I know you must have—anybody says they didn't think about suicide is a goddamn liar."

"Yeah," I said, "I remember. It's a wonder anyone ever survives."

The bus was late. Or I was early and anxious. It was cold out and everyone was talking snow. The woodpile was low and I decided that we'd try to go firewood cutting the next day before Brushy Butte got snowed in. Put him to work, give him something to do besides sitting around the house all day when he should have been in school. Give me something to do besides staring at the flickering cursor on my computer screen.

Edgar's worn-down clutch might blow out on the muddy landing and leave us stranded up there. But even if that happened and we had to hoof it four miles out to the county road, at least we'd be busy instead of just sitting around feeling sorry for ourselves. Besides, we really didn't have enough wood at home to stay warm through even a middling cold snap.

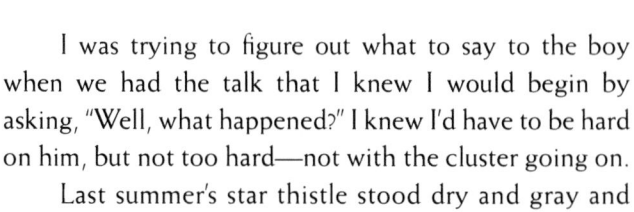

I was trying to figure out what to say to the boy when we had the talk that I knew I would begin by asking, "Well, what happened?" I knew I'd have to be hard on him, but not too hard—not with the cluster going on.

Last summer's star thistle stood dry and gray and popped-corn white along the railroad tracks. I could see the river running steelhead-fishing green through the bare tree branches. The South Umpqua turns that beautiful color when it settles down after a muddy brown winter rise and I always think of Chalchiuhtlicue, the Aztec goddess of earthly waters, whose name means Her-Skirt-Is-Made-of-Jade.

The bus came over the hump in the road, pulled up at old George's driveway, honked and backed up into it for the turnaround. The door swung open and Keith Gaynor's little grandson got out but he was the only one. I looked through my windshield at Betty, the driver, and saw her shrug and pick up a microphone.

"He didn't get on the bus," her amplified voice said, and she shrugged again and pulled out heading back toward town.

I remember sitting in the Highland Park police station waiting for my dad. I was fourteen then and my buddy Pat Miles and I had been arrested for shoplifting. A store clerk had chased us for three blocks and dragged us back to the store where we waited for the LAPD to haul us off in handcuffs. It was in June 1966, while my father was recovering from his second heart attack.

I remember how worried he looked when he came into the room and shook hands with the detective. Dad was only forty-six then but he was gray-haired from raising nine children and worn down by heart disease. On the way home I promised to stay out of trouble with the police. Three months later, just a few days before I started high school, my father died. A few years ago I finally stopped blaming myself for his death though I still regret having worried him.

I kept my promise for four years before I got arrested again, for vagrancy, in Arizona. My accomplice, Pat, was in and out of juvenile hall and drug rehabilitation programs during those years. He finally hung himself in a jail cell at eighteen years of age.

Where was he? Why didn't he get on the bus and come home? It wasn't supposed to be like this. At least, it never occurred to me that the lumpy-headed little bald guy who once lay in my lap would someday scare the hell out of me.

I followed the bus back to town, wanting to swing Edgar impatiently around the fat yellow rear end of the school bus. Instead, I lit a cigarette and ticked off the half-dozen places he might be, trying to replace my speculative visions with a search plan: check the school first, then the Super Y Market, then downtown to the park, the mini mart, Stan's Market, the baseball card shop, the Dairy Queen and then—what? Keep cruising, watch the sidewalks, stop any of his buddies I ran into and ask them if they'd seen him.

There was a basketball game going on in the gym and I spotted one of the half-wild Steltz boys up in the stands. I made my way past the scorers' table and he spotted me coming. "Hey Bob, what's up?" he called down to me.

"Hey Jerry, hey, you seen my boy?"

"Yeah, he's up at detention. Him and Jesse were late for Mrs. Fent's class."

"Thanks, man. I was wondering why he wasn't on the bus."

"Yeah, he's up there."

"OK, thanks."

I saw him sitting in a classroom with a half-dozen other boys and girls, all of them staring at the blank blackboard and watching the clock while a teacher sat correcting papers.

Twenty minutes later he came out with the rest of them. "Hi Dad," he said.

"Hi son. Detention, huh?"

"Yeah, I was late for class again."

"Well, I wondered why you wasn't on the bus, now I know."

"Yeah," he said, and we walked out to the truck.

He knew and I knew what was coming but neither of us said anything until I turned onto the old road along the river, toward the dump instead of crossing the bridge towards home.

"Where're we going?" he asked.

"Out to the rifle range," I told him, "I figured we'd take a walk down by the river."

"I was gonna wait 'til we got home."

I pulled off the road after a mile and crossed the ruts and puddles in front of the targets, rounded pebbles and shell casings crunching under Edgar's tires, down to the brush line along the river. We got out and walked to the bank and he sat down on a boulder. I picked up a rock and chucked it into the green water. "Well," I asked, "what happened?"

We were out on a muddy landing up on Brushy Butte the next afternoon, searching through last summer's logging slash for the good dry madrone branches that sound like baseball bats when you chuck them. It started to snow when we had about two thirds of a load, a few light flakes at first and then heavier showers, like gauzy curtains hiding the far side of the draw.

It built up on our hats and shoulders and the roof of the truck while we worked. The saw's chain was getting dull and the slash was turning white, making it hard to pick out the good wood in the tangle of logs and branches. I buzzed up one last chinquapin to top off the load and we loaded up the saw and maul and water jug, oil jug and gas can and jumped into the pickup cab.

I started the engine and while we waited for it to warm up he asked me one of those questions he likes to ask. "Why does God make snow, dad? I mean, what's the use of it?"

"Well, it don't run off quick like rain, soaks in good when it melts, so I guess that's one reason. But who knows? It's pretty, and He's an artist after all, so maybe He just likes to see the hills all covered with white."

"Yeah, maybe so. I never thought about that."

Hills and the heads of old people too, it occurred to me, He likes them white. That's why He gives us teenagers—it's a wonder any of us survive. But I kept that notion to myself.

"Well, let's see if we can get out of here without getting stuck in a blizzard."

"Come on, Edgar!" he urged the old truck, patting the dashboard, "You can do it. Go, Edgar, go!"

1992

# SNOW

The snow always comes as a surprise, even when you smell it coming and the weatherman tells you it's coming and everyone in the valley is talking about the snow coming and you see the sleet splattering on your windshield at sunset. Somehow you're never prepared for the reality of snow.

It comes so quietly in the night, in soggy clumps at first and then as tiny drifting flakes, each whirling crystal unlike any other that ever was or will be again. Snowflakes have the carefree innocence of all things brief and beautiful which fills us with sadness because they remind us of our own short stay here.

Of course, it's only us who feel that way. Watching the snowflakes fall, it's hard to imagine them carrying anything but joy in their whirling hearts. They are born in the bellies of clouds, tiny specks of earth's dust that clothe themselves in glittering sea moisture like debutantes dressing up in lace and jewels for a waltz on the winds. Believe me, though they never speak, they have their own soft song as they dance, singly or holding hands in the sky before they settle on the valley.

Even though we mutter about inconvenience as we make our way to the wood pile, or sit listening to icy road reports and school closures over a cup of coffee in the morning, our first and deepest reaction is delight. Like children bursting forth from the house to play, our hearts jump at the sight of the world transformed.

How strange it is to look out of windows edged in feathery crystal patterns of frost, to see sunlight split into rainbows by the icicle prisms dangling from the roof's eaves, to look out on fields of unbroken white snow.

The leafless fruit trees, oaks and maples so stark in winter rains reveal their delicate tracery when clothed in snow. Fence rails and posts, so common that they're almost invisible, stand out suddenly when piled high with white powder. How green and vulnerable the small patches of grass look, in sheltered nooks surrounded by the frozen snow. How warm and cozy it seems inside by the wood stove.

1991

# DEGREES OF UNDERSTANDING

It's always encouraging to hear from someone who admires your work. After all, freelance writing is a remarkably unrewarding career, a lonely and distant business where you send pieces off into the world like orphan babies to be adopted by utter strangers. You wonder what sort of homes they'll find, whether the child you bore through weeks or months of careful labor will be loved and cared for or neglected and abused.

You tuck them into a manila envelope pinned with a note that pleads for a chance to have a good home, and worry about their reception. Will they live on to support you and survive you and perhaps to help give birth to future generations? Or will they die prematurely, lying at the bottom of a cockatiel cage thousands of miles away? You drop them into the dark void of a Postal Service mailbox and all you can do is to hope that they have the strength to walk on their own.

I recently received a phone call from a local journalist who had read one of my essays in a literary magazine. She seemed quite excited by it, which was flattering of course, and I thanked her for her kind words.

I wallowed in the reporter's praise like a grinning dog rolling on a summer lawn as she described her chance encounter with my work in a Portland bookstore. She'd been struck by the insights my piece had given her and then, out of curiosity, checked the biographical notes in the back where she was amazed to find that we lived in the same small rural community.

She asked me a few questions about the essay and about my career which I answered easily but then she asked a tough one: "Is it true that you're a high school drop out?"

The question is always a little hard for me to deal with. Usually, the topic comes up less directly when I'm asked "So, where did you go to school?" It's one that used to embarrass me. Somehow, I guess, which college you've attended is supposed to be a measure of something—but I'm not sure just what.

"I'm a high school drop-out," I usually tell them. Strictly speaking, that's not the truth, though it's close enough for strangers. Besides, the truth is a bit complicated and it's a line that gets a satisfying reaction, so, why not fudge a bit?

Actually, Eagle Rock High School and I gave up on each other. "Fail to honor people and they fail to honor you," as Lao Tsu says. We were caught up in a game of mutual disrespect whose beginnings were lost but which ended when I loaded up a backpack and hitchhiked around the country for several months.

I was supposed to do well in school. My Scholastic Aptitude Test scores always ranked me in the top one percent. And yet, I had a tremendous antipathy toward schooling—perhaps a Scholastic Attitude Test would have been more useful.

Two conflicting views of why anyone would bother to learn were presented to me as a student. The first was that learning was a noble and joyful thing, worthy of pursuit for its own sake. The other was that poor grades spelled financial ruin. "If you don't do well in school, you'll never get a decent job," was the refrain. I could feel the truth of the first proposition because I enjoyed learning things regardless of their utility. But the second proposition smacked of coercion and I could never understand what grades had to do with the acquisition of knowledge.

There was something inherently false about schooling which I sensed but could not articulate. I saw it in the faces of students cramming for final exams and in the faces of teachers who wished they were elsewhere.

Learning has turned out to be a life-long joy and I never have managed to find a decent job, so I guess both propositions contained equal truth. I spent my first ten years out of school working thirty different jobs as a laborer.

After an on-the-job accident left me partially disabled, I found myself unable to labor full time and so I taught myself to write, a job with more prestige but considerably less income than laboring. I fell—ten feet, head-first—into it. A ladder twisted out from beneath me and, like Alice down the rabbit hole, I was pitched into a bewildering world of chronic pain and humiliation—the worker's compensation system.

For the first time in my life, I was unable to work. It is hard, after years of defining yourself through pride in your body's strength and endurance, to find yourself disabled. Poems are written about athletes whose careers are cut tragically short by injury or death but a laborer who gets crunched on the job is merely categorized as a "flake" and discarded.

I was told to get work pumping gasoline but I turned to my typewriter for solace and challenge instead. It was a new world to compete in but I felt sure that hard work and stubborn persistence would eventually pay off in my literary labors as it had in the woods and mills and construction sites. (Like so many notions in my life this one turned out to be correct but terribly naive.) Most of what I earn still comes from hiring myself out for the odd jobs that come my way.

⁂

Although I earned my General Education Diploma and even completed a freshman writing class at a community college, most of my notions about higher education come from dealing with college students and graduates. It's not fashionable to speak of class differences here in America, where we cherish the belief that we live in a classless society, yet, I admit, I am aware of the differences between myself and the academically educated.

My background has given me an understandable, perhaps unavoidable, belief that blue-collar workers are generally better people than white-collar workers and professionals. There's an element of sour grapes in that, to be sure, but truly, I do believe that working indoors stunts human beings just as a lack of sunlight stunts plants.

It's hard for me to trust people who haven't worked in the woods and the fields, in the mills and on construction sites. It's hard for me to trust people who have a college education and the prejudices which go with that—not that I am without prejudice myself, but mine are simply different. Even highly educated, sensitive, liberal people suffer from prejudices which are difficult to overcome because they are so deeply ingrained.

I remember arguing with an editor once, a very refined woman, holding that artistic sensibility was an inherent trait that crossed class lines. "Just because you're a poet or a dancer or a sculptor, that doesn't mean that you appreciate beauty any more than a logger or a field hand does," I told her, thinking about an old gyppo logger I knew. "Nonsense," she said, "of course it does."

I wondered at her ignorance and concluded that she couldn't see in others what she had never looked for in them. Elitism is a sign of ignorance, an ignorance which is the result of the blinding effects of prejudice, refined and cultured prejudice certainly, but prejudice none the less.

There is an outlook that laborers develop which is hard to explain to people who haven't done it. It comes from a life of pain, pride and desperation, from having to prove over and over again that you can keep up with anyone, and knowing that someday you won't be able to out-work the others.

We trust what we know. My knowledge is from the perceptions of my body rather than my mind. I can't believe in abstractions, the world of ideas, theories, propositions and labels where what something is called is more important than what it actually is. The world of mind seems dwarfed and dull to me when compared to the opulence of physical reality.

This physical education, the lessons which I carry in my body, my mind and psyche, is difficult to express and perhaps impossible to teach using words. What I've learned may not be the sort of things that lessons are made of (at least the kind of lessons that can be embodied in an essay). What do you learn from fifty-hour weeks of pulling and stacking lumber besides tolerance to noise, fatigue and pain?

Having worked for a number of college grads over the years, I found them to be oddly incompetent and

overly self-conscious. They seemed to be full of technical expertise yet utterly unfit for handling laborers. It
was as if they'd learned to substitute facts for fortitude
and tentative answers for tenacious actions.

I'm not ashamed of having earned five college
credits and a 4.0 GPA. I learned some useful things in
my brief time there, but I'm more proud of my years at
Sidewalk University's School of Hard Knocks where I
earned a BA (Busting Ass) degree and an MS (More of
the Same).

A friend of mine, who dropped out of the same
high school at the same time as I did, used to work as
a mechanic. A few years ago he had his own garage in
the small town where we both live. One day, he was in
the shop working on a pick-up truck and had his tape
player cranked up, listening to music while he worked.
He looked up from the engine to see a well-dressed
man standing in the open doorway with a puzzled
expression on his face.

"He was staring at me like I was a two-headed
monkey," my friend told me, "So I went over and said
'Hi! Can I help you?' The guy says, 'I was just passing
by and, uh, well . . . That's . . . that's J.S. Bach!' like
he couldn't believe it, you know, so I said, 'Yeah, it's
the Cologne Symphony doing the Third Brandenburg.
Heck of a nice recording ain't it?' And he just stood
there like he couldn't figure it out, you know, 'What the

hell's going on here?' The poor dude, it just blew his mind, I guess. I felt like I ought to put on some Merle Haggard just to make him feel at ease."

My friend was right to feel pity for his visitor, after all, nothing in the man's education had prepared him for this. Apparently, no one had ever told him that formal education can't create a love of great art or that a lack of formal education won't stop people from admiring it.

He'd learned instead a set of prejudices concerning the value of institutionalized education. He'd been led to believe that somehow the money he was spending and the time he was investing in academic studies was going to buy him a place among an elite group whose refined taste was their exclusive property.

Academics have an undeniable economic interest in promoting this brand of prejudice. Of course, this pecuniary interest is never a direct cause—that is really taking the charge too far—but it's undoubtedly an unconscious factor which hinders the questioning of the underlying assumptions.

I have another friend who earned a master's degree in creative writing and has never sold a piece to a paying market. He teaches English and uses some of my essays as models of good writing for his students to study. He finds it ironic, as I do, that over the many years of our friendship, he's made more money talking about those essays than I did by writing them, and that he earns a living as a certified master of a craft in which he has never done journeyman level work.

A university degree is not a reliable measure of sensitivity, talent, competence, knowledge or wisdom. It is, however, a sort of password among those who decide who to hire and who not to hire for certain positions. It says, "I am one of you. I share your outlook. I know how to play by the same rules you play by." Of course, the mechanic's overalls, the logger's suspenders and hickory shirt, and the trucker's baseball cap, say the same thing and serve the same purpose. Ignorance and prejudice are no one class's exclusive property, but then, neither are sensitivity, talent, competence, knowledge and wisdom.

1993

# PAINTING A HOUSE

*"Well-diggers lead the water wherever they like;*
*fletchers shape the arrow; carpenters bend a log of*
*wood; wise people fashion themselves."*
—THE SUTTA PITAKA CA. 250 B.C.

I was taught house paint-
ing by a friend who'd learned the trade from two old
men. As he instructed me in the rudiments of the craft,
I could hear their voices and attitudes coming through
him to me, an old fashioned way of working which
conjured up visions of medieval artisans raising cathe-
drals, for along with the mundane matters of scraping
and sanding there was an insistence on propriety.

Each instruction was accompanied by the phrase
"The professional painter," a standard which drew me
irresistibly to emulation like "the superior man" of Lao
Tsu. In every facet of the work, from cleaning the buckets to holding the brush, I found that there were many
ways to do it but only one way acceptable to a real
professional.

Gradually, I learned to discern the difference between a well-done job and an unskilled or careless one. Driving by a painting crew on the way to work, my friend would assess their level of professionalism in a matter of moments, by the way the painters were dressed, the manner of their preparations, the layout of their tools, the number, positioning and quality of their tarps, or the brand of paint they used. I found that the difference between a properly done job and shoddy work lay in the amount of attention paid to uncountable minute details, any one of which would seem inconsequential in itself.

Like life itself, a building being painted offers an almost endless series of decisions. Anyone can quickly color a house with new paint but how well it will stand up over time depends on the work you can't see, the care taken to prepare it before the paint goes on.

Every house has a painted record of the care and abuse that has been given it over the years. A good house painter can read it as easily as you would a comic book. The scars of use and the marks of time wear away at the paint, of course, and demand attention. But much of the work is concerned with the efforts of past painters, finding and correcting sins of omission: careless sanding, drips, overspray, thin spots and unpainted surfaces tucked away in obscure nooks, cheap paint— all the little "screw it" spots, where somebody, either through lack of effort or skill, didn't take the time to do it right.

House painting, like any craft, changes its practitioners. Somehow the rhythms of the work and the
habits it requires alter the painter. It's a slow process,
much like the work itself, a lot of small, step-by-step
minor changes, seemingly endless in the busy-doing
middle-of things, which can only be appreciated
afterwards.

Breaking a huge task down into a series of necessary steps teaches patience and hope—good train-
ing for attempting any lengthy labor, whether social
reform, writing a novel or having a fruitful marriage.
Anything worth doing takes time, discipline and attention to details. "The professional painter," my friend
told me, "is never in a hurry."

There is a steadiness that the discipline of repeating an orderly process brings. It begins on the physical
level and slowly shapes the mind and spirit as well. The
work transforms the worker, inevitably. Conscientiously
done work makes us conscious while careless work
makes unfeeling zombies of us. It also shapes our world
too, making it tidy or sloppy, beautiful or ugly, according to our attitude in the making of it.

A few years ago, while I was painting a big, rambling, old two-story house, I found myself ready to die.
I was finishing up the month-long job, painting the

trim on an attic dormer which sat too far back from the edge to be reached with the long aluminum ladder from the ground. It was dangerous, but no more dangerous than a lot of things I've done over the years working on houses, and I was being careful. As I carried my brush and paint bucket across the steep roof I realized clearly that I might die trying to paint that house. It was an odd feeling, not a suicidal or ominous one—just that I was ready.

The thought didn't hold any terror or grief for me and I found that I didn't have any regrets. I was simply ready to accept death. It was a very contented feeling really. Death didn't seem so bad after all and neither did life.

It lent a kind of grace to my work. Everything I did while filled with the acceptance of my own death took on a stateliness like a calming ceremony. Somehow, the work, the day and myself were stripped of pretension and clearly just were, without encumbrances.

❧

As with most manual labor, there's a lot of time spent in doing work that doesn't require much mental concentration. Chipping, scraping, sanding and washing walls keeps the hands busy and leaves time for reflection. Adolph Hitler painted houses for a living and I can easily imagine him in spattered white overalls mulling over the social problems of his time up on a ladder, just as Eric Hoffer pondered the nature of fanaticism while loading and unloading ships in San Francisco.

Of course, the work of preparing a house to be painted isn't really mindless at all, but in the early stages it is the choices which require attention more than the work itself. Later, as the coats of paint build up, there's a reversal. The choices become fewer and the execution becomes more demanding.

It is the final work, painting the trim, which demands the most careful concentration in the doing and the least thought as to how. The order is invariable, governed by a few simple rules: work the edges before the middle, the top before the bottom, the inside before the outside. What to do is easily said but how it's done makes or breaks the job.

Every painter keeps a special brush, called a sash tool, for painting window frames, door panels, and the fine details. A good painter babies this brush, cleaning it with great care and storing it lovingly after use. They vary in size and shape and materials, angle-cut or straight-cut, ox tail or boar bristle, and each painter will pick one type and use no other.

It's a wonderful, sensual feeling, laying the paint on, like petting a cat. To get the paint to lie just right takes a steady hand, a good eye and a physical attentiveness that keeps the mind from wandering, concentrating only on the flow of bristles across the wood. The long, even strokes that pull the paint to the edge without touching the glass must not waver. Too close, they spread over onto the glass; too far and the edge looks ragged with bits of the old paint showing.

Your body responds to the rhythms of the work, breathing in time to the long strokes that spread the paint evenly and, despite the slowness of movement, more quickly than short ones.

Body and mind and the work act on each other and shape and limit each other in complex ways. It's difficult to say just how and why, and every job is a special case, but in the end there's a result, an undeniable record of all that went into it.

***

I once took on a job painting a radiator shop as part of a barter deal. Lonesome Pick-up Bert, my geriatric 1948 Chevy, needed an expensive new radiator and the old metal building needed painting.

Decades of paint hung in great peeling, multicolored flakes from the place, giving it the sad look of a molting chicken. The landlord, I learned, was an old world craftsman called Herman the German, a man whose drive toward perfection had led him to spend seventeen years in the construction of a commercial building across the street. Herman's work had been studied and documented by a host of building inspectors, architects, engineers and professors who came to marvel over the workmanship of the place.

I thought of Herman the German constantly as we scraped and sandblasted. It was one of the oldest metal buildings of its kind in Los Angeles, one of the first gas stations in town, an ornate steel box full of odd angles, exposed nuts and bolts, tiny glass panes and decorative trim, built just before America's entrance into the First World War.

I was determined to make the tattered old bird rise like a bright Phoenix from the ashes of neglect, but I grew more and more dissatisfied as the enormity of the task unfolded itself in the work. As the expenses of time and money mounted, I found myself forced to compromise on my goal of perfection.

Near the end of the job, I was up on a ladder painting the thin iron window frames when I heard a thickly accented voice speaking to me from below.

"You boys are doing a very good chob," he said.

I turned to see an old man smiling up from below. The ancient master craftsman had finally stopped by to look at his rental. "Well," I offered lamely, "it could be better."

"Better?" laughed the old man, "*Ja, ja,* it could be better—always, it could be better. But you are doing a good chob. It is enough."

I was delighted by the praise, of course, and only later did I realize that he'd revealed a secret: Not perfect but good. Craftsmanship is a matter of compromise because imperfection is the nature of reality. "The professional painter" like "the superior man" knows this.

We paint our lives like this too, close—as close to perfection as the exigencies of time and material and energy allow—but never perfectly. To strive for perfection and to accept falling short gracefully is the way of sanity.

Knowing that achieving perfection is impossible brings humor and compassion into our lives by allowing us to take comfort in satisfaction, for living a life is a work too and comes to an end as every job does, and the result can only be judged by the effort and skill that went into it.

1991

## COUNTING HEADS

The first thing I had to do was to abandon the speech we'd rehearsed, which, if I recall it correctly, began with something like: "Hello, my name is (your name). I am an enumerator with the United States Department of Commerce (display your badge) and we are conducting the 1990 census in your neighborhood. With your permission, I would like to ask you a few questions regarding your household."

No one talks like that around here and it would be profoundly rude to adopt such a tone with my neighbors. Instead, the typical conversation began something like this:

"Mornin' Ma'am. How you doin'?"

"Why, just fine, thank you. How 'bout yourself?"

"Oh, pretty fair—can't complain. Looks like we're in for another hot one but I reckon it beats bein' out in the rain, anyway."

"Yeah, I s'pose it does. You the census man?"

"Yes ma'am. I hate to bother you like this, but you know how the government is, it's gotta get done. You got a minute?"

The government's notion was to be productive and professional, that is, impersonal. But you just can't ask people for personal information and walk out on them without leaving a little something in return. It didn't have to be much, maybe just an acknowledgment that my temporary employer would probably hand out the details of their personal lives to anyone who cared to take a look—if it didn't get lost first. More often, I simply needed to listen to their problems and share a few of my own.

All of this was time consuming, of course, and for a while I worried about spending too much time sipping coffee at kitchen tables. However, we had been instructed to "follow the local customs" and chewing the fat with the neighbors was demonstrably an integral part of our traditional local culture. Besides, I wouldn't have taken on the task of visiting seven hundred households if it hadn't been for the chance to hear people talk.

When I heard that the job was coming up, I knew immediately I wanted to apply for it. It was an instinctive response which I rationalized later, on the basis that meeting all my neighbors would be good for me as a writer. This was a reasonable and entirely justifiable excuse, but in the end, only an excuse.

The truth is that I, and every writer I've ever met, was a snoop long before taking up "my craft or sullen art." I've always loved eavesdropping on the conversations of strangers, watching their actions with a sidelong glance, secretly delighting in unguarded revelations.

Sometimes the craft is just an excuse for the habit, giving me the sort of professional immunity granted to priests and psychotherapists and census takers. At my more honest times I see that we writers are really a pitifully maladjusted bunch, hiding behind white sheets of paper, speaking with the great voice to hundreds of thousands of distant people we never meet while keeping our thoughts to ourselves among our neighbors.

The high point, topographically speaking, of my six weeks as an enumerator came when I stood on the bald top of Sheep Hill, the mountain at whose foot our little shack by the tracks sits. I'd been meaning to climb it for twelve years, but never got more than a third of the way up. Armed with the plastic badge of mandatory cooperation pinned to my shirt, I obtained the key to a locked gate from the telephone company and drove to the summit on a gravel road.

Typically, in this country of jumbled terrain, I had to drive ten miles to reach a point about a mile, for those with raven wings, from my front door. It might as well have been Tierra del Fuego or Fargo. To hike it would probably require a tough six hours round trip, because half of that mile-long shortest distance between two points is up.

My official purpose was served the moment I reached the end of the road just short of the microwave transmission tower. I could say with certainty that no one lived up there without ever leaving the car. But I couldn't turn away without spending a federally funded idle hour enjoying the kind of vista that once served vision quests and is now reserved for utility workers.

Standing on a mountaintop changes your perspective, not just optically but, more importantly, emotionally. The few square miles of valley nestled between ridge tops which makes up my day-to-day world seemed at once both grander and more fragile than I'd imagined. It was as if I stood in two places at once, up above and down below, both laughing Gulliver and self-important Lilliputian.

Gathering information has hazards more dangerous than walking up to a neighbor's porch past ill-tempered dogs. Perhaps the greatest danger was in seeing the data but not the people. For one thing, none of the really important questions appeared on the survey forms. I began to doubt the validity of a statistical abstract in forming a portrait of my community. In the real world, no matter what arithmetic you use, the answer is always the same because, in the end, there's only one of each.

As I went from door to door, methodically circling clockwise and always working the homes to my right, I began to toy with the notion of a meaningful survey, devoid of names, gender, age and occupation but full of the really big questions and with room for answers that might require volumes just to record a single reply.

At every sixth home I conducted an in-depth interview, using a long questionnaire requiring highly personal information about income, jobs, level of education, military service and health. I was often stuck by the contrast between one long-form interview and the next. One elderly couple, living in an old wooden house on a $6,000 per year pension, reported ten years-worth of elementary school education between the husband and wife. Six houses down the road, a couple in their late twenties were earning over $100,000 annually. Twelve homes up the road, a farmer with a college degree reported working more than sixty hours per week during the previous year for a net loss of $40,000.

On the same mile of road, I met a bartender who lived in a trailer, husbandless and raising two children. She was in a hurry and asked me to come by her establishment that evening. It was a slow night at the bar with a half-dozen regular customers lined up on the bar stools. I ordered a cup of coffee and we worked the interview in between her duties. One besotted old harpy, sitting three seats down, kept offering advice to the young woman, repeating the same defiant mantra with each question, "Name, address and Social Security number, Honey—that's all you gotta give 'em."

Actually, none of the forms asked for anyone's Social Security number, but the long form did require me to ask for her job title and then a description of her main tasks and duties.

"Mixing and serving drinks," I suggested and began to pencil in my own reply anticipating her agreement.

"Nope—babysitting drunks," she informed me in all seriousness. I turned my pencil upside down to erase what I'd written and recorded her reply.

Even in its smaller elements (in this case Douglas County, Oregon voting precinct 63A) a society resembles an untidy rummage sale where cultural artifacts, ranging from cobwebby attitudes left over from days of obsidian tools to this morning's radio talk show topic, crowd together on the same cluttered table. One old farmer, who managed to withhold every bit of information without ever refusing to answer a single question, gave me a rambling account of his family's struggle against government perfidy going back to the days of the Greenback-Labor Party and William Jennings Bryan's Cross of Gold speech.

The joy of discovering traces of the past was one of the first of many unexpected delights I found and it nearly became an obstacle to the others. To see what is actually there to be seen is harder than digging up stumps. Our perceptions are so deeply rooted in what we expect to see that we almost never notice what is truly there.

I was determined to understand my neighbors, not just to stockpile ore for the stamping mill of my craft, but out of a sense of social and spiritual duty as well. We are, as members of our society, required to know our community in order to make the informed decisions which, as citizens, we must make. Knowing the issues and ourselves is not enough; we have to know each other as well.

Coupled with the biblical mandate that we love our neighbors, the obligation to know the people we live among becomes an essential matter. It's not merely an ideal goal to work toward but, in the most fundamental sense, it is an inescapable necessity. Even from the most pragmatic view point, we simply can't live an enjoyable life unless we know, and through knowing, love each other.

Looking back on my stint as an information gatherer, the strongest impression I have is of the nearly universal decency of my neighbors. Although many people took the opportunity to complain about the government, and a few refused to cooperate, they were polite about it. Well, almost all of them were polite.

At an isolated trailer house a nervous young couple answered the door and explained that they were just visiting and that their friend had only recently moved into the place. In accordance with my instructions to base the survey on residency as of April 1st, 1990 I asked them if they knew the exact date. Before they could reply, a loud male voice began booming, "Wrong! Wrong! Wrong!" from somewhere inside. They looked at each other and then at me in alarm. Then the chanting gave way to a stream of expletives the drift of which sent me scurrying to my car.

I paused a moment in the car to note the address and check off the appropriate box on the census form before leaving. A bearded, shirtless, skeletal, yellow-skinned, scabrous fellow appeared in the doorway and accused me of writing down the license plate number of a van parked in the driveway. A clinical diagnostic term, "amphetamine psychosis," and the possibility of gunfire occurred to me simultaneously.

"You got a problem?" the emaciated fellow demanded to know. I did, but I took care of it by starting my engine and jamming the transmission into reverse.

Still, other than a speed freak, a bar fly and a few toothsome dogs, no one barked at me and, even then, I never got bitten.

---

After a few weeks, the wearing effects of concentrated attention finally caught up to me. The sheer numbers of people and places and conversations I experienced day after day became more than I could keep track of. It was then that I became aware of something I'd felt before but never really appreciated fully—the benefit of being burned out.

When the tide of statistical information and personal observation reached flood stage it was sink or swim, so I floated instead. I didn't choose to adopt a different attitude. I was simply too overwhelmed to do anything but enjoy what I was doing without adding any layers of moral or intellectual involvement. Oddly, giving up the attempt at understanding allowed me to know more than I'd thought possible.

There are times in each of our lives when for one reason or another we simply can't spare the energy it takes to prop up our self-image. This is a very frightening condition sometimes, but on other occasions it is a source of great joy. Either way, it is always a time when we are able to learn unexpected things because we're too tired to do anything but directly experience unadulterated reality.

With my attempt at noting all the differences among my neighbors a hopeless mess, I finally began to appreciate their underlying sameness. In an unexpected way, by looking for the little things that set them apart from each other, I'd been ignoring a large part of who they were.

Observations, especially when it comes to people, always tell us more about the observer than the observed. Every conversation is a Rorschach test. Until you have the knack of losing focus, or, more properly, of abandoning your perspective, it's impossible to appreciate the multiplicity of an ordinary human. When you simply accept people without attempting to define them you learn more about them, even in the briefest of encounters. With people, face value is full value.

---

A friend of mine, who holds a doctorate in sociology, shuddered when I pointed out to him that categories of people don't really exist. "My God," he moaned, "but that's all we do. Our whole science is based on breaking people up into categories so that we can compare them to other groups."

I felt bad about that, after all, he's good guy and terribly concerned about finding ways for people to get along with each other. But, there's always more to people than we can understand. People are real but definitions are always false, or at best useless without a deeper understanding.

---

I've listened to the earnest talk of people who are struggling to understand a notion called "community" and wondered what particular bit of earth, if any, they hold dear. Like each individual, each community defies definition, though not understanding.

Robert Frost understood. His poem, *The Death of the Hired Hand*, probably comes as close as anything I've read to the truth about community. In the poem, an old homeless drifter, sick and nearing death, returns to a farm where he'd worked for many years as a hired man. The farmer and his wife take him in—he grudgingly, she compassionately. In the course of a good-natured but earnest argument over the derelict's unexpected arrival and what they ought to do about it they try to define the word "home."

"Home is the place where, when you have to go there, they have to take you in," he snorts.

"I should have called it something you don't have to deserve," she replies.

I find myself returning to this place which I have not left. Often, in some distant conversation about right and wrong, or love and hate, or the struggle to define what is or isn't, my thoughts come back to these few square miles of land and to the people who walk the same ground in their daily rounds as I do. It has become more important to me than I can say to know that we are all here, together.

1992

FOR JUSTICE

Nobody gets it and it's hard to explain, so I really got a kick out of Fair Share Bob's Myrtle Creek story.

My wife and I were in Veneta for the Oregon Country Fair, Lane County's annual summer counter-culture extravaganza. The Emerald Empire, Eugene and its environs, has a reputation for being Oregon's most liberal, politically correct community but the Country Fair, even by Eugene standards, is a leftist event.

We were in Community Village, a little cul-de-sac of haphazard wooden booths housing dozens of radical activist organizations. The purest of the pure—goddess worshiping New Age neo-pagans, Earth First!ers, pro-hemp lobbyists, no-nukers, eco-feminists and organic farmers—gather there every summer to congregate, celebrate and preach to the faithful. And here was Bob from Oregon Fair Share, an advocacy group for the poor, all excited about having done door to door canvasing in Myrtle Creek, my town.

"It was the weirdest thing," he began.

(Uh-huh.)

"I was a little worried about how they'd react to us anyway because it's supposed to be such a redneck area . . ."

(Among the politically correct, Douglas County, Oregon, is known as one of the most hopelessly incorrect places in the state.)

". . . but the stats say it's really got some major economic problems . . ."

(At that time, a few years ago, they would have read something like: fourteen percent unemployment, seventeen percent poverty rate, nineteen percent emergency food usage, twenty percent adult functional illiteracy.)

". . . and so we felt it was important to canvass down there and get the word out about Fair Share. You know, see what's really going on and all."

(Uh-huh.)

"So, we're coming into town and the first thing I see is that weird sign about the vigilantes."

WELCOME
our streets
Patrolled by
MYRTLE CREEK'S
VIGILANTES for JUSTICE

"And I'm thinking, 'Oh my God! What are we getting into?' So we go over to the city hall, because we have to register under the Green Valley Ordinance before we can go door-to-door, and they're all like, 'Well, OK, you can do it, but be careful—no telling what might happen.' So the first place we go to there's this pick-up in the driveway with a gun-rack and a bumper-sticker, says 'Old F.A.R.T.—Fathers Against Radical Teenagers' and now I'm really getting paranoid, you know?"

(Stephen King. Rod Serling. "Easy Rider." "Deliverance." Rednecks with shotguns and pick-ups—Oh my!)

"And this guy invites us in and it turns out he's one of the Vigilantes and we sit and have coffee and everything, talked to us for an hour. We couldn't get out of there; they just kept talking and giving us cookies and all. They were so nice. It turns out the Vigilantes are just a bunch of old guys with CB's, a Neighborhood Watch kind of thing."

"Everywhere we went it was like that. We'd planned to just spend a day in town but it took us three days because everyone treated us so well. I couldn't believe it. Up here in Eugene they'll slam the door in your face sometimes but nobody was rude down there—they didn't give us any money—but everybody was just so friendly. Nicest bunch of people I ever met."

I laughed when he finished, but not out of derision, for him or the town. I felt instead an overwhelming sense of pleasure, the kind of feeling you get when, after years of toying with a complex set of vague notions and contradictory facts, all of your half-realized conclusions are suddenly confirmed at once. Just about everything you would ever learn about Myrtle Creek was contained in his little anecdote. I could have kissed him, but I couldn't stop laughing.

***

Everything they say about Myrtle Creek is true—sort of. They (our urban neighbors on the Willamette and Rogue) say it's a backwater place, the domain of ignorant bible-thumping, "God, Guns and Guts" hillbillies.

They (the folks who live here) say it's a friendly, progressive little town, combining the best of old fashioned community values and modern development, a real nice place to raise your kids.

And they're both right—sort of. Besides, the truth is a little complicated and not nearly as believable, or as much fun, as the myths.

Take, for instance, the large wooden sign that greets you as you cross the bridge into town. The Chamber put it up, turning a weedy patch of hillside into a landscaped greeting scene. "WELCOME TO MYRTLE CREEK, GATEWAY TO THE ONE HUNDRED VALLEYS OF THE UMPQUA" it says, and hanging underneath they've added a white one this year, "1893 Centennial 1993"

Setting aside the fact that no one has ever counted just how many valleys there are in the 5,000 mountainous square miles of Douglas County, it's hard to see Myrtle Creek as the gateway to much of anywhere. It's located a good thirty miles north of the Josephine County line and Interstate Five bypasses the town over on the other side of the river. You have to go out of your way to get into town and the only two valleys you have to pass through the town to get to are North Myrtle Creek and South Myrtle Creek.

Also, there's a controversial footnote to the centennial, since the state revoked Myrtle Creek's city charter for a couple of years back in 1901 due to the town's failure to collect enough taxes to operate. Some people argue that the two years without a charter makes it only ninety-eight years old and others that the second charter was a new one and therefore the current town only goes back to 1903.

So, in all honesty, the sign should read something like: "WELCOME TO MYTLE CREEK, GATEWAY TO TWO OF THE ONE HUNDRED (more or less, but nobody's really sure how many) VALLEYS OF THE UMPQUA. 1893 (or 1903) Centennial (except maybe for two years when the town was bankrupt) 1993 (or 1995 or 2003 depending on who you ask)." As I said, the truth's not nearly as believable as the myths.

In Myrtle Creek boosterism reaches almost pathological proportions. The town hosts an incredible number of festivals, concerts, and amateur sports contests ranging from Babe Ruth Baseball state championships to wheelchair basketball fundraisers.

This flurry of public-spirited display has increased over the past decade or so as the town made painful adjustments to a series of economic disasters. The slump began with the Reagan recession of the early eighties, continued with the "lean and mean" (when bosses got mean and workers got lean.) "miracle" recovery and—ever downward—slid into the Bush era of housing slumps and environmental litigation.

One early casualty was the annual Wood 'n Nickel Days celebration which had been held for more than twenty years. When the town's last mill shut down and Hanna Mining Company closed the Nickel Mountain strip mine and smelter in nearby Riddle, the celebration, in a town now lacking both lumber and nickel ore, was rechristened the Summer Arts Festival. Considering the times, maybe it should have been called the Unemployment Check and Government Surplus Cheese Festival, since those had become the mainstays of the local economy.

Desperate times call for desperate measures. After all, a town cannot live by Food Stamps alone. So, when a local restaurateur suggested an annual race to help liven things up, the business community took up the challenge.

Most people don't know this, but, by crossing two pieces of lath and laying the X across its back and running fast enough to keep up, you can guide a panic-stricken turkey more or less in a straight line. The hugely successful Turkey Grand Prix brought out happy crowds to watch local business leaders, dressed in appropriate costumes, sprint down Main Street behind live fowl.

It was great fun, but after a couple of years some animal-rights activists got upset and threatened to spoil the event with picket signs calling media attention to the "cruel and barbaric ritual." Despite the general consensus that anyone who feels sorry for a turkey must have never raised one, the Chamber caved in and switched to pushing frozen turkeys down the street in wheelbarrows and shopping carts. Political correctness had come to Myrtle Creek at last, but it just wasn't the same and the Turkey Grand Prix died out for lack of interest.

Millsite Park is another product of the Hard Times. In the late seventies work began to convert an industrial waste land, the former site of a mill that burned down in the fifties into a city park.

The park, a volunteer project which is still underway, is a remarkable achievement for a town of 3,300 residents. So it's true that the old fashioned barn-raising spirit of neighborliness is alive and well here.

Which is not surprising because the people aren't too far removed—only a generation or so—from rural self-sufficiency.

City dwellers tend to forget that the word "neighbor" (from the German *nach bauer* by way of the Old English *neahgebur* meaning "near farmer") is a rural word for a rural concept. "Civilization," on the other hand, comes from the Latin *cives* meaning "city." The two ideals are not really related. A place can be, and often is, neighborly without being terribly civilized and the reverse holds true too. In fact, civility and neighborliness just might be notions as incompatible as urbanity and boorishness.

And that, in a nutshell, seems to be the source of the confusion about Myrtle Creek, both in its image outside the area and in its image of itself. Like other small rural towns all across the American West, Myrtle Creek is being carried headlong into the twenty-first century and making the painful transition from neighborliness to civility.

For two hundred years, country folk all over the world, under the relentless pressure of industrialization, have been forced from the self-sufficiency of the land into the wage dependency of the cities. Cut off, literally, from their roots, within a generation or two they lose their rural values, which were based on cooperation and broad common needs, and exchange them for the market ideal of narrow competitive goals. And there, as they say, goes the neighborhood.

The ancient bond of the nach-bauerhood, already virtually extinct in American cities, lives on in Myrtle Creek, at least for now. But here too it's starting to show wear and tear under the strains of civilization. The hard times of the eighties gave us both the renewed cooperative spirit that built a beautiful park and a new generalized atmosphere of betrayal, distrust, and fear.

Whenever there's private fear and desperation, there's always someone ready to focus it outward on a wider, less personal, unsolvable problem. After all, it's always easier to worry about something like "family values" than to think about what will happen to your family if you can't make your next land payment. And once people get in the habit, it's easy to carry them along from one fear to the next.

Law and order, predictably, was the first one. The Vigilantes for Justice came out of the "get tough on crime" and "war on drugs" years. Next, yellow ribbon fever struck and the town erupted with anti-environmentalist posters, and, of course, yet another sign at the edge of town proclaiming Myrtle Creek a "Yellow Ribbon City" where "We Support the Timber Industry."

During the past five years, support has come to mean opposition and with the cornucopia of targets, a siege mentality has set in and increasingly complex conspiracy theories go the rounds as the issues all become somehow intertwined. It seems, sometimes, that everybody here is actively opposed to something. Actually though, what it amounts to is a sizable minority that is opposed to just about everything from Earth First! to the school board.

---

Despite the hard knocks of the past thirteen years and the changing nature of life in this valley, apathy, at least, is not much of a problem. Myrtle Creek is still a small enough town to provide its inhabitants with a sense of responsibility and the power to determine their own fate. This is a rare attitude nowadays, one that survives only in backwater places, like a rare flower pushed to the brink of extinction by the destruction of its natural habitat, which is a shame, because it is the essential ingredient for the type of democracy Thomas Jefferson saw as the great hope for America's future.

1993

A DANGER TO THE COMMUNITY

### ROADSIDE TRASH

The old guy wasn't as old as he'd looked standing there all crumpled and discarded by the interstate with his thumb out. Once he got in the truck I could read "wino" written in ravage lines on his face, as plain as if it'd been block printed with a black marker. He had furtive dumpster-picker eyes, half sly fox and half beaten dog. He also had a mild case of the shakes, either habitual tremors or from alcohol withdrawal. It's hard to tell which unless you ask, which is a rude thing to do.

I regretted stopping to pick him up even before he sat down next to me filling my pick-up truck cab with his rummy smell. Picking up hitchhiking derelicts is a habit I can't seem to shake—probably a result of listening to too many nuns' stories about Jesus or Saint Peter begging at the door in disguise. The poor bastards are harmless, though sometimes unpleasant, company.

Besides, I've never gotten the truly frightening, creepy feeling from a wino that I get regularly whenever I'm around glad-handing realtors or bankers or politicians.

"I ain't going far," I told him, "just down to Myrtle Creek."

"Well, every bit helps," he sighed, "I'm heading to Medford. How far is that? I'm not too far am I?"

"No, it'll be about seventy miles from where I drop you off is all. With any luck you'll get there in an hour or so."

"Good," he said, "I'm in kind of a hurry. I want to get into a detox program. I heard there's one down there they might let me in. I tried here in Roseburg but they said it'd take me six months, 'cause I ain't got no money to pay for it."

That explained the shaking. I hoped he wouldn't start his delirium tremens before I dropped him off. A hardcore alchy could die from sudden withdrawal without medical treatment.

"Yeah, well, that figures," I told him, "the place is packed with people who got busted on piss tests at work. They're too busy raking in the insurance money right now to mess with you." It was the spring of 1988 and random workplace urinalysis testing was a multi-million dollar front in the latest election year War on Drugs.

"Take a look at this," he suggested and rolled up his shirt sleeve to reveal an impressive looking crater in the crook of his arm.

"Did you show them that?" I wondered aloud.

"Yeah. Didn't do me no good though."

"Been waiting long?" I asked to change the subject.

"About four hours, I guess. Not bad, about average."

At four hundred cars per hour, the average traffic flow on that stretch of road, he'd watched 1,600 drivers pass him by.

"Yeah, that's not bad at all. I picked up a pair of Canadians there a couple weeks back. They'd been stuck for fourteen hours."

"Jay-zus!"

"Yeah, they was bummed, couldn't believe it. They thought folks around here must be extra unfriendly or something. The fact is that people are just scared. A hitchhiker murdered a local gal here a couple years back so nobody wants to give anybody a ride."

"Well, I can't say as I blame 'em. That's all it takes, one bad apple to screw it up for everybody else."

"Yeah, it's been in the papers for a long time, real brutal. Turned out the guy'd escaped from prison and she picked him up on the freeway. The guy raped her, cut her throat and stole her pick-up. They caught him and there was a big trial and everything."

"Oh, Christ, that's sick. I mean, I can understand— a guy gets a little horny and what the hell, you know? But, shit-oh-dear, you don't gotta go and do something like that," he shook his head. "You know, we had a guy down in California years ago, raped a fifteen-year-old girl and chopped her arms off. Left her for dead."

"Lawrence Singleton."

"Yeah! That's right, Singleton—that was the guy's name. How'd you know that?"

"He's out on parole now. The papers are all full of it. Nobody wants him to live near them and he can't find a place to live. He might be moving up here from California pretty soon."

"Here?"

"Yeah, he's got an invite from this preacher down in Azalea, about twenty miles south. The folks around here are pretty upset about it. There's going to be a meeting Saturday down at the Grange."

"God, I can see why. Who'd want a guy like that around?"

I dropped him off on the interstate on a spring morning six years ago, but, in a metaphorical sense, I guess I'm still hauling him with me. I wonder sometimes whether he lived long enough to get the hospital bed and intravenous rehydration, tranquilizers and food and vitamins he obviously needed. I hope so. I hope that God really does have a special love for drunks and kids and crazy people, and that he didn't just end up an anonymous corpse somewhere.

I think of him sometimes when I listen to the vague jargon of activists and politicians, people who speak in metaphors so much that it seems they never consider reality itself. The old guy "fell through the cracks in the system" in poli-soc jargon, though if he actually fell, he more than likely fell on some linoleum or concrete or

asphalt, leaving an untidy John Doe cadaver for someone to haul off to the morgue, rather than dropping
neatly down a metaphorical chasm in something as
abstract as a system.

I never met Lawrence Singleton, alcoholic and
parolee, but I've talked to many unwanted people.
They're not hard to find in a country like ours, where
lives are tossed away as casually as cigarette butts.

## GOD'S COUNTRY

My friend Esther called for some advice, or some help
anyway. She'd been talking to one of the reporters
who were asking everyone up and down the road questions about the Reverend Tom Smith and the Brides of
Christ. It wasn't easy to explain why Smith and his flock
were so unpopular with their neighbors and how that
distrust had erupted into a national news story. Esther
thought that I should talk to some of the television and
print reporters who were suddenly covering Azalea like
maple bugs on a window screen.

"He was asking me about Smith's dog getting
shot," she said, "and I tried to tell him about how it
works around here if your dog gets into the neighbors'
livestock but he didn't understand. They're from back
east, city people. How can you tell them it's okay to
shoot a dog if it's killing your chickens?"

I imagined Esther trying to explain the complex
etiquette of shooting your neighbor's dog to a journalist

from back east. The reporter was probably accustomed to liquor store shootings, rapes and muggings, but the notion of using a 30-06 on a trespassing mutt would strike him as petty barbarism rather than an unpleasant necessity sanctioned by law and custom.

Dogs, either on their own or in packs, will chase, maim and, often, kill livestock. Sheep and poultry are particularly vulnerable, but calves and colts are sometimes mauled. A pair of dogs out on a lark can destroy thousands of dollars-worth of sheep in a brief frenzy. Ranchers and farmers, faced with a marauding canine, shoot. If they miss, the county Animal Control officer will track down the offender and send it to the shelter for destruction and write out a heavy fine for the dog's owner. The proper thing to do when your dog is shot by a neighbor is to apologize and pay for the damaged livestock and count yourself lucky to avoid the fine and court costs.

I could see how that might be tough to explain, at least in any form that would fit into a sound bite. I could also understand why the preacher and his flock might interpret the dog's untimely demise as harassment. After all, they really were an unpopular group, as the bullet holes in their bus and front porch proved.

When the Reverend Tom Smith bought half of Marmalade Farm from a friend of mine, he didn't mention that he'd be bringing his sixty-five member flock to settle there. Nor did he say anything about other land purchases he was making, a home on Starveout

Creek, another property downstream on Cow Creek
and a commercial building in Canyonville. The neighbors were more than a bit alarmed when a seemingly
well-financed group of men, women and children from
Las Vegas appeared suddenly in their valley.

Azalea, Oregon, is not a town, just an old wood
frame general store with gas pumps and a post office,
zip code 97410, serving about 250 homes. The people
of Azalea live scattered along a twenty-five mile stretch
of Upper Cow Creek Road, which runs along a shoestring valley from I-5, where the store is located, into
the mountains. It is one of hundreds of small valleys
tucked between the tangled mountain ridges of southern Oregon, as isolated from each other as islands in
the sea.

Although it has no city charter, the place comes as
close as anywhere in America does these days to being
a community under the old-fashioned definition of a
place where the residents know each other. Out in the
boonies what counts is not politics but neighborliness,
the willingness to help each other simply because of
proximity.

When the Brides of Christ showed up in
Canyonville people called them the "hankie-heads"
because the group's women invariably wore bandan-
nas tied over their hair, a style which they had report-
edly adopted due to some obscure biblical injunction.
Seeing them at the Pioneer Super-Save or the Sears
Mail Order Store reminded me of the sharp-eyed old

Russian women who used to live in North Dakota back in the fifties, though the *babushkas* favored long silk scarves tied under their chins rather than cotton kerchiefs tied at the nape of the neck. The look was a bit eccentric but appealing in a nostalgic sort of way.

They were a clannish bunch, not given to socializing much, which is bound to raise speculation. There was their "kiss of Christ" custom, for instance—some kind of an *agape* thing—which they practiced and with it the rumor that the women were brides of Reverend Smith as well as of Jesus. It was said that too many of the group's many children bore an uncanny resemblance to their spiritual and temporal leader. Was the kerchief a badge of servitude rather than modesty?

When a local paper revealed that Pastor Smith had a criminal record which included a conviction for child molestation, folks took it as an affirmation of their suspicions and secret fantasies, proof that there was hanky-panky going on among the hankie-heads. Smith granted the local daily paper an interview in which he freely admitted his pre-salvation depravities and offered his sordid past as proof of God's great love for even the lowest of sinners. He also wondered if perhaps bigotry wasn't the real reason why he and his multi-racial flock weren't getting along with the neighbors. Still, Jesus had warned that His followers would be persecuted, so it was to be expected.

An uneasy truce set in between the true believers and the community. The Brides were absolutely law-abiding and undeniably industrious. People began to assume a more tolerant attitude. After all, Southern Oregon has a long history of eccentricity, ranging from nineteenth century utopian experiments to the countercultural communes of the seventies.

The Reverend began to mail out a monthly newsletter to local residents in an effort to promote understanding. Unfortunately, he succeeded. In a remarkable series of epistles, he laid out his convoluted teachings, drawing connections between apocalyptic biblical prophecy and a conspiracy theory linking Catholicism, Freemasonry, the Bavarian Illuminati, the Trilateral Commission, and the federal government (which is headquartered, it seems, in a city laid out in the form of a gigantic Satanic talisman). Only a chosen few, Christ's brides, would survive the time of tribulations which the Whore of Babylon was about to unleash from the power vector of evil located inside the DC beltway.

Though the notion of a diabolical presence in the Reagan administration had a certain appeal, it was just too tidy and comforting. It would be heartening to find that there really is a purpose behind "the evil that men do" since that would be a sign of competence and intelligence.

Inviting Lawrence Singleton, former axe-wielding rapist, to live out his days in the bosom of the one true church, was a very Christian thing to do. Certainly, even if nobody else wanted him, Jesus did, which is the nice thing about Jesus of Nazareth, equal opportunity savior. The Lord never has been too particular about the company He keeps. Love them all—let God sort them out.

## THE KLIEG LIGHT CRUSADERS

The Reverend Smith didn't display much political savvy, but this lack was made up at the Grange Hall meeting by an out-of-district U.S. congressman, two state assemblymen and a half-dozen county office holders—people who knew how to conduct themselves on a flag-draped podium and weren't uncomfortable in front of television cameras.

The theme was crime and the speakers were vehemently, almost violently, opposed to it—certain forms of it anyway. They didn't mention the crimes which had been committed against the Brides of Christ who, while they'd outraged the community, hadn't broken any laws. But the crimes of Lawrence Singleton were condemned in suitably lurid terms. Although the possibility of personal forgiveness and redemption wasn't brought up, salvation for the entire community was offered through a crusade against crime.

The atmosphere was that of a revival tent. As the politicians denounced criminals and molly-coddling liberals equally, the audience grew palpably more righteously enthusiastic. The expressions of uneasiness and concern on their faces gave way to a wolf-like eager grin showing confidence in the justness of their wrath.

By the time congressman Denny Smith, the featured speaker, came to the podium to lay out his plans for mandatory sentencing, expeditiously applied capital punishment and the building of camps "with Quonset huts and barbed wire and machine gun towers" over in eastern Oregon, I half-expected him to lead an angry lynch mob in a march up to Marmalade Farm. Instead, he ended his speech, as had the others with assurances that he would aid the community in ridding themselves of the cult which threatened them.

As it turned out, Congressman Smith went away after the klieg lights were shut off and neither he nor the national press ever returned. Lawrence Singleton announced that he was moving to Florida rather than Oregon. The Brides of Christ packed up and left for eastern Washington a few years later. In the end, the only threat to the community turned out to be the secular cult of law and order and the demagogues who, for a few short weeks, exploited it.

1994

AT THE ZOO

Just recently, here in my own home county, someone splattered a front window of the local Democratic Party Headquarters with a bucket-full of feces. Behind the glass was a life-sized cardboard cut-out picture of President Barack Obama, who is an unpopular man in this neck of the Northwest woods for many reasons. It is impossible to know what motivated the unknown poop-flinger without a confession or claim of responsibility of some sort, but three likely possibilities come to mind: vandalism, political discontent and racism.

Like any downtown section of any city, Roseburg has on-going problems with vandalism. These, however, have always been the sort of things one might expect from teenage hooligans or drunken stew-bums. There has also been, over the years, some prankish vandalism at both the Republican and Democratic Party offices—crazy-glue in the door locks and the like—but nothing before this has ever made such a splash.

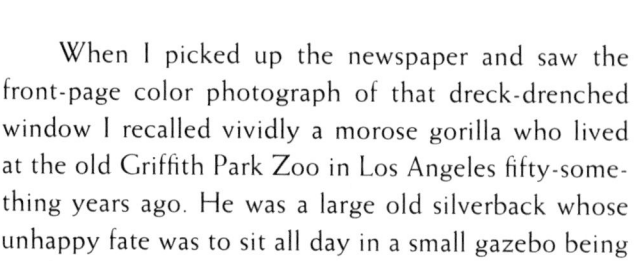

When I picked up the newspaper and saw the front-page color photograph of that dreck-drenched window I recalled vividly a morose gorilla who lived at the old Griffith Park Zoo in Los Angeles fifty-something years ago. He was a large old silverback whose unhappy fate was to sit all day in a small gazebo being gawked at by people.

The people encircled his small round steel cage completely and there was always a crowd. Since there is both proverbially and actually "always one in every crowd" there was usually a yahoo, or two or more, trying to provoke a reaction from the sullen beast by waving their arms, shouting and flicking cigarette butts at him. The gorilla, understandably, was discontented with his lot and expressed his malcontent with sad sour looks and by flinging feces against the glass barriers that surrounded his enclosure.

The ape, of course, had it in his nature to act as he did. Some fear-driven instinctual reaction to his unnaturally stressful life caused him to seek relief in the form which gorillas (and other great apes and monkeys) do. Since (most of the time anyway) humans don't fling excrement at each other, this—how we deal with our fear—is a difference between us, the humans, and our near genetic relations, the apes.

We humans are blessed with two great advantages in the form of superior intelligence and superior culture when compared to the apes. For most of my life I believed that the apes, at best, had a very limited intelligence and, therefore no culture at all. Nowadays it is known that chimpanzees and gorillas are at least as smart as an average four-year-old human—smart-enough to develop culture.

Chimps, it turns out, do have a sort of culture. They learn things, imitate and invent much as we do. So, what keeps them from writing bad poetry and charging each other interest on loans? In short, why are they (so nearly identical to us genetically) chimps and why are we human? Many scientists now believe that the answer lies not so much in our brains as in our adrenal glands.

Chimps, as any zookeeper will tell you, can't stand to be around each other for very long. They scream at each other and menace each other, fueled by high levels of adrenalin, the "fight or flight" hormone. They are, by nature, extremely fearful and distrustful of each other—very much like human meth addicts, constantly on the alert.

It is our human ability to remain calm in each other's presence that allows us to create civilizations and their apish inability to do so which condemns them to short brutish lives in the brush. Call it love, call it trust, either way it seems, it is in fact we humans, the meekest of apes, who have inherited the earth. It is only the trusting and loving who are humane. To trust no one is, perhaps, worse than insanity—it may be atavistic, apish, less than fully human.

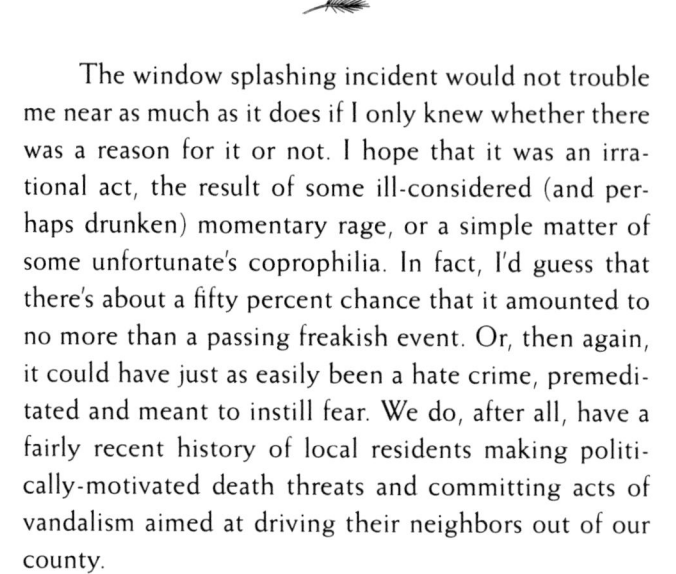

The window splashing incident would not trouble me near as much as it does if I only knew whether there was a reason for it or not. I hope that it was an irrational act, the result of some ill-considered (and perhaps drunken) momentary rage, or a simple matter of some unfortunate's coprophilia. In fact, I'd guess that there's about a fifty percent chance that it amounted to no more than a passing freakish event. Or, then again, it could have just as easily been a hate crime, premeditated and meant to instill fear. We do, after all, have a fairly recent history of local residents making politically-motivated death threats and committing acts of vandalism aimed at driving their neighbors out of our county.

"Highly conservative" is the usual description of the voters of Douglas County, Oregon and it is true that we Umpquans generally vote at a rate of 2 to 1 for Republican Party candidates and at the same rate against nearly every tax increase, no matter how laudable its purpose may be. But it has been an unusually anger-filled year here in Douglas County and in the nation itself as well. There were bitter "Tea Party" protests here in the spring and, this summer, so-called congressional "town hall" meetings which turned downright ugly and hateful at times.

Some of my friends and neighbors ask me, "Who are these people? Why are they so rude and so angry?" Much too distressingly often, they put it to me as, "How could they be so stupid?" without realizing just how arrogant and ignorant that question is. Others of my friends and neighbors are among the very people being asked about and, they are neither less nor more intelligent than the others.

I suppose that by "stupid" people are asking about the all-too-human willingness to believe outrageous lies and specious theories. I think perhaps naiveté might better (and more kindly) describe this tendency of people to believe in dubious (but comforting) half-truths and fictions—yet it too comes up short. Brain power has nothing to with it and neither does education. I have known many highly-intelligent people who have this very same affliction and it is one that is present in every economic class and every nation and all races and both sexes, taking root in the uneducated and the highly-educated alike. Intolerance, I think, is not really so much a matter of opinion as it is a symptom of underlying psychological problems—an indication of character flaws which are tied to unresolved anger, a generalized lack of trust and an inability to remain calm when facing life's ambiguities. There is always an inevitable percentage of humanity which is distrustful, easily excitable and emotionally unstable.

I have known a great many people over the years—nice people, decent people—yet people who cling to harmful and repugnant beliefs which are racist, homophobic, xenophobic, misogynistic or politically intolerant. What they all have had in common is their high levels of frustration and fear. Each has felt insecure and cheated somehow, denied their fair share of power, ignored and disrespected. Many (though not all) have been economic losers, bitter about their failure to succeed. Some have been emotional cripples, unable to sustain loving relationships and unable to tolerate ambiguity. Many have had their lives fall apart due to compulsive boozing or drug abuse or gambling. Others have simply been crushed repeatedly by an indifferent and impersonal system of things which exploits them because it is profitable to do so. Some are people who blame themselves for having suffered terrible blows which came for no good reason at all. All became, in one way or another, shell-shocked veterans of life itself.

What is there to cling to when, by your own doing or by others or by cold fate, you have lost everything? Stripped of dignity, mired in failure, caged in by tough circumstances and uncontrollable forces, what is left to people but to embrace comforting nonsense and to rage against perceived injustice?

A while back I ran across a fellow who wanted to know where Pomerania was located. We were both in a local bookstore at the time and when a woman entered the store carrying a small mostly-white lap-dog he inquired loudly, "Where is Pomerania anyway?"

I explained to him that "Pomerania" was the name of a region located along the Baltic Sea coast in what is now northwestern Poland but which used to be north-eastern Germany. "G'dansk is the biggest city there. It used to be called Danzig," I added.

At the mention of the old port city at the mouth of the Vistula his eyes lit up. "Danzig," he confidently informed me, "was named for the Tribe of Dan—one of the Lost Tribes of Israel." He went on to describe how this particular Hebrew tribe, which had been missing in action since Biblical times, had left the Promised Land to wander into Europe and left their name scattered across the face of the continent which they populated. His list of Lost Hebrew Tribe of Dan place-names included any name in any European language which featured a "d" and an "n" separated by any vowel: Danzig, Denmark, Scandinavia, London, Sardinia, the rivers Danube, Don, D'neister and D'nieper had all been allegedly visited by these ancient Jewish name-leavers.

It seemed to me a peculiar misunderstanding of both European history and of the way languages work. But he was so clearly pleased with his display of erudition that I didn't have the heart to tell him just how absurd what he was saying actually was. Besides, the conversation had taken place in the Religious Books section of the store and clearly these bizarre notions

were somehow tied into his spiritual beliefs. Since it
is one of the oddities of human nature that irrational
beliefs are perversely reinforced by factual challenges
to their validity I gave him up as an interesting nut-case.

Later, a quick internet-search for the term "Tribe
of Dan" brought the matter into a more disturbing light.
To begin with, I found that there actually is a contem-
porary Tribe of Dan but that no one suspects them of
having been lost Hebrews since they are all black folks
who live in West Africa.

What I did discover is that his rap is a popular one
with an anti-Semitic White Power religious sect known
as Identity Christians, the modern adherents to a nine-
teenth-century crackpot movement known as British
Israelitism. Among these believers it is an article of religious faith that Jehovah, having cursed and abandoned
the Jews of the Holy Land, allowed his Chosen People
designation to fall entirely upon the descendants of
the Jewish (but blameless for the execution of Jesus
of Nazareth) Tribe of Dan, who were the ancestors of
today's modern Christian Europeans, particularly the
fairer-skinned inhabitants of the continent's northern
nations. Oregon, it seems, is currently home to two
congregations of this odd persuasion, one located in
Woodburn and the other in Eugene.

❦

That there are malcontented people in our society
is not surprising. Things being as they are—theoretically

egalitarian but factually equal only in our shared vulnerability to chance personal disaster—it is inevitable that some of us will have happier lives than others and that the discontented will resent the happiness of the contented and will view them with envy and distrust. This is sufficient evil in itself and creates, on its own, a good deal of trouble.

There are those in our society though, who, seeing discontent and the fear behind it, want to use that fear and resentment to further their own ends. Just as the sight of the gorilla moping in his cage brought some people to pity the unhappy beast, some to scorn it and others to taunt it for their own amusement, so seeing the fears of the down-trodden brings some to compassion, and others to contemptuous exploitation.

A few years ago I ran into an old high school buddy of mine while I was down in Los Angeles working the tradeshow circuit. The town, I'd noticed, had changed in the decades since I'd left, not just physically, but culturally as well. Racism had, once again, become nearly main-stream thinking among the white citizenry. There was much grousing about the increasing numbers of Armenians, Asians and Hispanics living in the over-crowded county, a strong demographic fear of being overwhelmed by swarthy foreigners whose large families were allegedly bankrupting governmental resources.It was a pleasure to run into my old school chum again, who, it turned out, was working as a city Parks & Recreation gardener at the new Griffith Park Zoo. He invited me to visit him at his work for a private early-morning behind-the-scenes tour. The new Los Angeles Zoo is a much nicer place than the old one. The animals on display have more room to move about and their enclosures are designed to mimic natural conditions. We have come a long way, I saw, in learning how to reduce stress levels in captive wild animals.

My old pal, however, seemed to have changed for the worse. Once an open-hearted and unafraid artist, he too, I found, had taken to seeing himself as an oppressed member of an endangered soon-to-be minority of white people. It disturbed me to hear him talk so vehemently about his anger and frustration over the political and social concerns of the moment. I wondered at it, at the time, and later learned that he habitually listened to Talk Radio programming while raking leaves and planting flower beds.

I'm not sure why so many people whom I've met over the past decade or so get such a kick out of listening to a steady stream of bad news and outrageous commentary. I suppose, once again, it may come back to the adrenal glands. I think of the thrill-seeking of my youth and the exciting rush that rose within me with the risk of physical harm. Anger, too, brings on much the same hormonal waves. I've heard the term "politics junkie" used to describe my friend's habit and suspect that it may be more of a reality than merely a metaphor.

It saddens me to see my neighbors deceived. I don't blame them much though. Lying to people for profit has become a multi-billion dollar international industry. I condemn instead those who have deceived them, the professional liars and, even more so, those who employ the liars.

Those who profit from the subversion of reason, who inflame smoldering anger for personal or ideological gain, who appeal to the worst in human nature—to our anger, distrust, resentment and greed—are much more dangerous to the "domestic tranquility" of this nation than the majority of common criminals. A thief, a burglar, or a robber only harms a few victims but those who spread anger-inducing lies may harm millions and their harm often out-lives them and perhaps circulates for centuries.

"It is the first duty of the humanist and the fundamental task of intelligence to ensure knowledge and understanding among men," according to Pablo Neruda, a man who certainly knew about such things. Of the demagogues, professional gasbags, spin-doctors and liars-for-hire who prey upon the vulnerable I would ask, "If the old poet was right, then what is it to ensure ignorance and misunderstanding among people but to be working against humanity?"

Looking back on it, it was an odd return to the zoo, a place I hadn't been since childhood. It seems strange to me now that I returned, after fifty years, to find that the animals are calmer and the people more resentful, frustrated and angry. What would have happened, I wonder, if we'd spent our time and money coming up with ways to reduce our human worries instead of coming up with ways to increase our anxiety? We have, since Neolithic times, made tremendous advances intellectually and technologically but we have not advanced psychologically at all. We are still going about the world with our frightened caveman hearts which are increasingly ill-adapted to worrisome distractions and stimuli that would have been unimaginable just one hundred years ago.

We seem, as a society, to have a great deal of trouble in learning to forgive each other. We suffer the ill effects of a sort of Karmic footprint, rather like a Carbon footprint. It is said that it takes one hundred years for a pound of carbon dioxide to dissipate from the atmosphere. How long does it take for a hatred to no longer circulate? In some cases it can take a lifetime— for some cases, centuries. My ancestors often used to say, "Forgiveness is the best revenge." I have returned to that saying many times over the years and always found it to be true. It tells me that by reacting to injury with anger or violence I am damaging myself more than I am my enemy. It really is better to forgive and forget and to leave the fear and anger and frustration to the harmful to bear than to carry those festering wounds within. This is what the best of humanity's teachers have taught since ancient times.

2010

DIRTY LAUNDRY

I was talking to Bob the Milkman in the laundromat in town. It's a spacious, comfortable, old run-down place, of peeling paint and broken linoleum tiles, with worn-out machines that cough and spit and shake like epileptics. It was the kind of conversation that might have taken place in a coffee house or small cafe, where people gather to talk earnestly. But I'm not sure whether such places even exist anymore and, even if they do, our little town doesn't have one.

"Well, I see the rate of entropy is increasing rapidly," he began, and he wasn't talking about the laundromat. We started out with the elections just a few days past and moved quickly on to politics, economics, apathy, alienation, television and what passes these days for civilization—everything that has the stench of death or worse still, the sterile, inhuman lack of any smell at all.

We sat, two aging hippies, engaged in the old eternal, foolish task of trying to find a meaning to it all, to somehow make sense of the incomprehensible. Here among the dirty laundry, on a side street in a small town we searched for a way to reverse entropy.

Bob was feeling even more cynical than usual. It was obvious to him that the world was rapidly heading for hell in a laundry cart. Of course, though he professed despair, he hadn't given up looking for solutions. It takes a lot of love to be a cynic; you have to care before you can be bitter.

"How?" he kept asking, "How can it be overcome? The technology is so pervasive. Television, for example, serves the status quo and people have forgotten how to question the conditions that keep them powerless and apart—and one of the main ones is TV itself."

I'd been stewing in the juices of my own personal despair when he came in with his broken wicker baskets filled with clothes, so it did my heart good to leave my clothes in the dryer after they were done and take up the defense of the good old cause for a while. It's amazing how the cleansing action of social optimism can scrub the embarrassing stains from your soul.

We talked of the impending war, of the savings and loan scandal, of the power of advertising in subverting reason and of the loss of hope in the daily struggle to pursue happiness through the endless swamp of major and minor restrictions to liberty.

It was a sort of game, like playing checkers with social principles, his black negative pieces to my red positives. Sitting on a pair of wooden benches, scarred with the jackknife-engraved history of our town's romances, we faced each other over the board of life, finding friendly pleasure in something we both took quite seriously.

He opened with the deception and hypocrisy of politics.

I countered with the wisdom of common people.

He didn't doubt the wisdom but what about apathy, the sense of powerlessness?

"Lack of power breeds frustration, a powerful emotion that brings change," I maintained.

"Only if it's a shared frustration, otherwise it just increases separation and distrust leaving us ripe for manipulation by demagogues."

"People know, deep down, that they need each other. Eventually people realize that things like dignity and justice and freedom are real needs—things we can't live without. If they're denied us, we'll get together and demand them."

That one shook him some, I could see it in his expression as he searched for a response.

"But who are we anyway? Where is our sense of common purpose, of ourselves as a people? People have to have something in common in order to band together. We've been reduced to seeing ourselves as individuals, cut off from each other, mere consumers with conflicting self-interests. There has to be some common ground."

"Well, there's always the sense of place, the realization that the common ground is right here beneath our feet. We are, in fact, neighbors here."

"That's an antique notion. People don't live in a place anymore."

"Whether they realize it or not, they do. It can't be denied because it's a physical fact. We share this valley. That's reality and reality is awfully stubborn—it keeps coming up no matter how much we ignore it."

"It comes up, but what if we don't have the ability to see it anymore? Our language has become debased by the system. It's so damn pervasive. We can't even think clearly anymore because words have become meaningless from being twisted to serve the status quo. All we have left is a vague, generalized discontent and no tools to examine it with. Whatever happened to the search for meaning? It's died out. People used talk about this stuff all the time. It was important to them. What happened to all that?"

Well, he had me there. I thought of the world I came of age in, the late sixties and early seventies and how tough it is to explain those times to teenagers. The kids see the shell of those days, long hair and tie dyes, rock and roll and LSD but it's incredibly hard to tell them about our idealism, our search for basic human values and our commitment to passé notions like truth and justice and peace and love. I was struck by how odd we two were, as anachronistic as two old white-bearded Jews discussing one of the fine points of Kabbalah.

"Yes, it's pretty much gone now," I admitted, "but not entirely. There's always going to be the misfits, the oddballs like us, who sense something's wrong and go looking for the answers. It will always crop up in little pockets here and there and—who knows?—maybe some of those seeds will take off and grow."

"Yes, a few seeds will grow, but they'll be stunted. The ground's gone barren. The soil's been sterilized. Hell, people don't even read books anymore."

"Yeah, most of my friends don't read books at all, and the ones that do, read garbage. The only people I know who actually read literature are all writers. It's discouraging. It haunts me. I don't know, maybe we need to find ways around that—storytelling or something. Maybe if we can find the right kind of manure the seeds will have a chance."

"Maybe, but don't count on it."

There wasn't much more we could say. He'd shaken my optimism and I'd rattled his pessimism and, strange as it may seem, we were both happier for it.

Outside, one of our town's shattered shell-shocked military veterans wandered by under gray November skies, wrapped in his personal cloud, searching for cigarette butts, lost in a forgotten war. Who knows? Maybe Bob's right. Maybe we're all becoming shell-shocked, on the road to becoming lost souls, searching for a happiness fix on the sidewalks of life.

I gathered up the warm clean clothes, loaded them in my battered pick-up and drove the autumn roads home, searching for the right kind of manure.

1990

## FALLING ASHES

The sunset shadow of the mountain creeps across the bottom lands toward the river. It's a late October afternoon in the valley, almost Halloween, and along the river the trees of autumn are bright in the sun. The strip of trees along the river forms a leaf-mosaic of gold, green, yellow and brown and even from a distance each leaf is distinct, out-lined in shadow. The salmon-road river quietly flows through the bottoms, low but on the rise, cleaning out the summer algae, waiting for the seasonal creeks to appear.

Across the river a peach orchard in red and yellow lies surrounded by flat fields, squares of green sheep-dotted pasture, hay fields and wheat stubble. Orange pumpkins lie in their bed of frost-bitten vines ready for the knives that will give them baleful eyes and jack-o'lantern grins. The setting sun shows the snake braid course of the old river channels, still visible a century after the soil-building marsh was drained.

The distant black figures of humpbacked men move along the edge of a brown field. Their out-stretched arms taper into bronze rods that drip molten lava. A line of flames marches across the stubble, crack-ling hungry flame-tongues lapping up the harvest chaff, making the soil ready for another crop of winter wheat. The smoke of autumn rises, making a reddish golden glow in the setting sun's light. The hills and mountains are distant seeming, dark and haze hidden.

Behind the smoke, Interstate 5 crosses the bottom-land as final and unswerving as an exercise in geometry. Beneath the crackling sound of the fire the highway adds the softened thrum of diesel engines and a rub-bery shush of eighteen wheeled semi-trucks. Like ants bearing the harvest, they follow each other in a busy endless line stretching north and south.

Close at hand, leaves pirouette in their brief moments of freedom and land with a soft crackling. Jays call raucously, quail and finches rustle in the poison oak brush in search of seeds. A wide-eyed calf gallops across the pasture to the safety of his mother's side. The old dog stares out across the valley, nose pointed toward the smoke, whining nervously as she smells the scent of the death of an Indian summer afternoon.

1991

## ROTOTILLING RESURRECTION

"*Spring has now unwrapped the flowers, day is fast arriving*
*Life in all her growing powers toward the light is striving*
*Gone the iron touch of cold, winter time and frost time*
*Seedlings working through the mould now make up for lost time.*

*Herbs and plants that winter-long slumbered at their leisure*
*Now bestirring green and strong, finding growth and pleasure*
*All the world with beauty fills, gold and green enhancing*
*Flowers make glee among the hills and set the meadows dancing.*"

— THE CAROL OF THE FLOWERS
16TH CENTURY, TRADITIONAL

Waltzing my tiller, I wore
a blister in the palm of my hand. I've gotten soft, apparently, sitting around all winter with the rain and snow outside my window, long days spent shifting a flickering cursor from line to line on my computer's cathode ray tube. It's hard to accept soft hands that, even with canvas gloves on, blister from work.

I used to take pride in my scarred and calloused hands and swore that they'd never be soft. But it's been years since I bucked hay or peeled poles or set chokers. A little firewood cutting to keep us warm, some yard work once in a while and perhaps a bit of shingling or carpentry once or twice a year is about the extent of my laboring now.

The scars remain but the protective calluses are gone. My hands are as soft as a butcher's or, worse yet, as soft as an editor's. I swore that I'd never lose touch with the gritty, hard and lumpy world of physical reality. I would stand, ankle deep to my rubber boots in manure, shovel in hand and tell myself how much easier the real stuff is to deal with than the metaphorical kind. And yet, I've gone soft now, become a presser of buttons and a peddler of metaphors. I fight it, but it's difficult not to become what you despise.

Rototilling helps. Everyone ought to go to bed tired and wake up hungry, of course, but exercise done for its own sake has always struck me as a sign of decadence. The sight of an exercise machine in the corner makes me shudder with visions of caged hamsters running on squeaky wheels. Besides, it's hard for me to understand why anyone would work up a sweat and not get paid for it.

It may seem odd, but spending a few hours operating a roaring, clattering machine that's constantly bouncing and vibrating beneath my hands and raising clouds of dust is soothing work. Although it requires an acute level of physical attention, (otherwise you could easily amputate a foot) it's not mentally demanding at all.

Spring weather is playful and being out there in the shifting sunlight, under a blue and white tie-dyed sky, feet sinking into the earth and head in the clouds, is a wonderfully expansive feeling after a winter spent cloistered indoors. The earth reveals its hidden secrets, the colors and smells of the varying soils. Quartz, agate and jasper, bones, old bricks, colored glass and pottery shards and silverware, lost marbles, and toy soldiers, all come popping up revealed by the whirling tines.

Every spring I take out an ad in our local weekly and hire myself out as a tillerman. My customers are mostly old folks who have a small garden patch to prepare and not enough energy to dig it up by hand anymore. For some, the back yard gardens are an essential source for food they wouldn't be able to afford. For others it's just something to help relieve the tedium of television and medication and visits to the doctor.

Rich or poor (and mostly they're very poor) they all need their little bit of freshly turned soil. It's hard for them, in a world that no longer needs their labor, to find a way to apply their useful skills. Talking with them about the weather and mulch and compost over coffee and cookies during my vernal visits, I get the impression that battling weeds is really their way of battling inertia and death.

In my first few tilling seasons, I naively imagined that it would be good to work for free. Of course, I couldn't (and still can't) afford to do that. And yet, it seemed a noble notion, to do a good turn for these impoverished old people, to perhaps pay back a little of the respect that life had cheated them of in their "golden years."

When we talk about soils and fertilizers and varieties of vegetables, age is not part of it. I am simply the hired man and we talk about the work at hand. There is no condescension or pity involved. Our roles are clear, employer and employee, and we respect each other on the basis of our knowledge and labor and skills.

Of course, they need me to prepare their ground. But lately I've begun to understand that they need me, as much or perhaps more, to sit and listen to them as gardeners, and that my wage, which they can't afford to pay and which I can't afford to decline, is the only way we can honestly honor each other. It is the precisely calculated measure of our respect, for ourselves and for each other.

1992

## BLACK WINGS

Visitors to the Umpqua Valleys are always a little disappointed when they find out that the big soaring birds they've been admiring aren't eagles after all but turkey vultures—what we call buzzards here.

"Oh . . ." they say, "Buzzards huh?" and then they change the subject, embarrassed at having mistaken a common carrion eater for the Lord of the Skies.

You can't blame them really, it has to be a let-down when you find that the bold, sharp-eyed hunter you thought you saw turns out to be someone who dines on road kill possum. They suspect that somehow they've been taken in by a shabby impostor and even if you don't laugh outright they know they've made the kind of mistake that country people enjoy seeing their city friends blunder into.

Of course, we never tell them how right their first instincts were, that we often pause to watch them ourselves, because their spiral soaring fills us with delight.

It's a local secret, something we don't often mention to each other—let alone admit to outsiders—but buzzards are beautiful.

Spring time comes to the valleys on black wings when the first buzzards return from their winter vacations in the deserts of Arizona and California. They come in from on high, tiny specks riding the wind, dropping lower and lower to sail along the rock faces and cliffs, wobbling like kites as they spread out over the valleys.

On sunny mornings after a rainy night they sit in snags and spread their wings, drying their feathers. They're sociable birds, patrolling with their friends and relatives or just roosting together five or six to a tree. They don't seem to mind sharing a meal. There's plenty to go around and besides, in their line of work I'd imagine one can't be too proud or fussy.

Though the buzzards migrate, they spend the better part of the year, nine months, right here, breeding and nesting and raising their young just like we do. Like good country folk everywhere they don't ask for much, just a chance to live quietly and peaceably among their friends and neighbors, getting by on what the land has to offer.

In late fall, when the cold comes and the valleys cool, no longer sending up thermal drafts for them to ride, they gather together, fifty or sixty in a flock, waiting for the right wind to take them south. One morning you wake up and they're gone and the skies are suddenly emptier and more lonely.

1991

Years ago, I lived in Los Angeles without a car or money for bus rides. I hitchhiked daily and spent hours beside the rivers of cars fishing for a ride. As I peered at the faces behind the windshields, I became fascinated by the expressions of the drivers. At first, I was struck by the great variety of people and later by the habitual masks they wore when no one was there to watch.

Faces tell us many things about people. The history of our lives is etched in lines of hope and fear, joy and sadness, caring and indifference for all to plainly see. Our passing emotions leave permanent traces at the corners of our eyes and mouths and as we grow older the sum of our inner life becomes more and more evident to everyone we meet.

Younger people can seem neutral in their look, preoccupied or bored perhaps, but with almost no visible hint of their outlook in their aspect. Old people though, seemed either happy or sad when alone and not thinking of anything in particular. It's as if life itself, that master gardener, had patiently shaped them like potted bonsai. I remember pointing this out to my wife one afternoon as we stood at a traffic light on Eagle Rock Boulevard, in northeast Los Angeles. We were very young then, she sixteen and I eighteen, and we made a pact to be happy-faced in our old age.

Half a lifetime after those sidewalk days in Los Angeles I'm still watching faces and as my appreciation for just how difficult life really is has grown, my fascination with the beauty of elderly men and women has grown too.

---

My younger brother calls them "grimsters," short for "grim-faced old men with caps" and you see a lot of them around here, driving by in their weathered pickup trucks, dressed in hickory shirts, feed caps and overalls or Lee 88's with suspenders. Their faces are as worn as old work gloves. They're retired loggers, farmers, miners and mill workers mostly, and most people just call them old timers.

Visitors from the crowded warrens of asphalt and florescent lighting where most Americans live often mistake the look for unfriendliness or bitterness. They remember television shows and movies featuring unsuspecting tourists caught in sinister small towns where violence lurks behind the hard sidelong glances of lounging rednecks, and grow uneasy.

People fear what they don't understand, of course. To me, bankers and accountants are scary people and the thought of a Chamber of Commerce luncheon fills me with a nameless horror and dread.

The old men my brother classified by their grim expressions are almost always kind, gentle and full of humor and mischief when you listen to them talk. Though they have their share of regrets, I never get a sense of bitterness or frustration from them. Their faces record a life of orneriness, determination, patience and resignation.

I spent two summers working for McCormick Piling Company in Riddle, Oregon, hand-peeling bark from logs destined to become power line poles. It was the last pole yard in the state where machinery hadn't taken over the job and it closed down four or five years later during the Reagan "trickle down" recession.

Pole peelers are now as extinct as the grizzly bears who used to live in these mountain valleys. Maybe it's all for the best. I'm not sure I really need half-ton predators who run the 100-meter dash in the low sixes ambling around my pasture and I can't honestly say that I wish I was still chipping bark from sixty-foot logs at eight cents per linear foot for a living. But still, something is missing here, gone forever and the valley somehow isn't the same.

It was fairly hard work and if you didn't have the right attitude and technique it could be brutal. Only old men could last long at it. High school kids, attracted by the notion of doing piece work with no time clock to punch, would show up and beg for a load of logs to peel. None of them ever lasted long. Most would knock out half a load and give up. But the four regular peelers, men in their sixties and seventies, kept at it day after day, wielding their double-bit axes, spud bars and Peavy poles at a steady, unhurried pace that the impatient young athletes couldn't match.

I was used to keeping up. Laboring is very competitive and I took pride in being able to work right alongside, and sometimes out-work, men my age and younger who were four inches taller and forty pounds heavier than me. After six years on construction sites and in the woods I considered myself as physically and mentally tough as anyone I'd ever worked with. But every time I paused to straighten my back and wipe the sweat off my eyelids I'd look over at the old timers and they'd be ahead of me.

The hardest lessons to learn are the ones you think you already know. I'd always used my head and my heart to beat men whose size and strength I couldn't match. Big guys rely on strength, little guys use technique, pacing and endurance.

I was always looking for ways to make my work quicker and easier, handling heavy materials with good body mechanics, laying out my work efficiently, pacing myself for the long run and simply resolving to endure whatever pain and fatigue the work week demanded. It took me a month to understand that I was being beaten at my own game by grizzled masters who could work a young buck like me to death.

As the weeks wore on I came to know them. We'd help each other free poles that were unusually large or jammed in the deck. Slowly, in laconic bits and pieces of advice they taught me the tricks of their anachronistic craft. "Let me see yer spud bar. Not bad, gotta good weight and the curve's about right, but see here? Your edge's sharpened up on ya. Gotta dull it some or it'll bite through the bark instead of slipping under." A bastard file would appear from a back pocket, three smooth swipes across the edge and it would come back to me. "Here, try 'er now. That spring steel gets sharper when you use it."

As the summer wore on our conversations got longer and further removed from the task at hand. They knew a lot, those old men, and I found out how to do many things that modern technology had made obsolete: how to sharpen the old two-man misery-whip saws, how to load piling on railroad flatcars with mules, how to hand-hew a tree trunk to make an octagonal sailing ship's mast ninety feet tall.

Not so long ago, most Americans lived in the countryside and even those who lived in cities made their living as blue collar workers. My grandfather, Lorenz Heilman, still farmed with draft horses in the forties and early fifties as he had done on his father's farm in czarist Russia before the turn of the century. People tend to think of those times, with a mixture of romantic sentimentalism and condescension, as a simpler time.

Nothing could be less accurate than to say those times were simple. Getting by in the old days, from "Whan Adam dalfe and Eve span" to 1950, required intimate, complex and sophisticated knowledge. My grandfather was illiterate but he could read a team of eight draft horses as easily as I scan newspaper headlines.

What we yearn for in the past is not simplicity but the certainty and self-assurance of those times, something which has died out in our culture within a single generation. Our lives here in the Information Age have become as temporary and transitory as the flickering cursor on a computer monitor. We have lost the sure knowledge of our bodies, the physical education that literally kept us in daily touch with the world around us. Henry Miller, back in the 1940's, wrote of *The Air-Conditioned Nightmare*, which the American Dream was becoming. Today, that nightmare has become our daily reality and we dream of "simpler" times.

The old timers know that their world has passed. They know, full well, that it was a brutal and ignorant time in many ways. One and all they believe in the inevitability of progress and think it good. The hope of a better life and the pride in helping it come about sustained them through hard and bitter times, through deaths of friends in war and at work, through the despair of economic slumps, labor strikes and false promises, through a life of pain and labor. They have arrived at last, tough and gnarled as ancient oaks, in the promised land only to find themselves in a world which doesn't seem to need them, where the old virtues of endurance, self-reliance, orneriness and integrity don't count for much.

"You know," one of my fellow pole peelers confessed one day, "I don't need the money I make here. I've got a pension and Social Security and all that. My land's all paid for and I could take it easy, but I've worked so damn hard for so long, it's the only thing I know. If I stopped working I'd just up and die."

A few years ago two elderly women from Siskiyou County, about 120 miles south of the Umpqua in Northern California, were driving from Yreka to Crescent City on a two-lane mountain road when their car went over the bank and down a steep hillside through the brush.

Both of them were injured in the crash and because of the rugged remoteness of the place they spent two days helping each other crawl back up the slope to the road where they flagged down a passing trucker who took them to a hospital. When a news reporter asked them how they managed to survive the ordeal one of them replied simply, "We are women and we are strong."

I laughed when I read that because I have a neighbor who undoubtedly would have done and said the same thing if it had happened to her.

Alice is in her late eighties or maybe early nineties now and she lives alone in a rocky canyon about a quarter of a mile up the road from me. I'm her nearest neighbor and over the years I've watched her place for her during her solitary trips throughout the western United States, when she drives the mountain and desert roads collecting rare wildflowers to plant in her rock garden.

Years ago she asked me to keep an eye on her place for a few months. At eighty-two she was having her knees replaced with artificial joints because her arthritis was keeping her from gardening. For three months, I stopped by every day and fed the dog and cat and kept an eye out for vandals, until one day I spotted her pickup truck sitting in the driveway.

I walked up to the house and saw her crutches lying by the toolshed but she was nowhere around. I called her name and walked around looking for her, worried that something might have happened. Visions of her lying on the ground helpless filled me with increasing fear as I searched, circling outward from the house.

Finally, I heard her calling my name from up the canyon so I set off up the steep trail along the creek. I found her carefully picking her way down the path, leaning on a shovel in one hand and a pick in the other, her knees still wrapped in bandages from the operation.

"Been up cleaning out my spring box," she explained, "Got to get the water working again so I'll be ready for planting. My garden club's coming out next month for a tour. You know any teenagers I can hire that ain't too lazy to put in a good day's work?"

1992

WITH A HUMAN FACE

A half-dozen years ago I was sitting around an evening campfire at the Oregon Country Fair in Veneta, just jawing with a few of the folks on my volunteer crew when, for some forgotten reason, I mentioned the Mudsharks in the course of my rambling.

"What's a Mudshark?" someone asked.

"The Mudsharks were a Hoedad crew," I told them, thinking that would be explanation enough for anybody, until the next question hit me.

"What's a Hoedad?"

I looked around at the half-dozen faces in the flickering light and saw innocent curiosity. They weren't kidding, I realized. They really, truly had no clue about who or what a Hoedad was. How was that possible? Was I not in Lane County, Oregon? They seemed so young then and I suddenly felt incredibly old and anachronistic, like some grizzled old-timer in a cheesy children's play croaking, "Gather round, my darlin's and I will tell thee tales from the ancient days of yore."

"The Hoedads were a bunch of hippie tree-planters," I told them after had I recovered from my shock and dismay, "they had a co-op and they used to be huge around here."

—⋇—

There was always something a bit tongue-in-cheek about the mighty muddy Hoedads despite their dead-earnest approach to worker-ownership. They were a strange bunch—leftist radicals and "simple life" hippies with reforestation contracts that, over the years, amounted to several millions of dollars. But, considering that they planted trees for a living, the strangest part was that a very high percentage of the Hoedads had college degrees.

Planting trees is the sort of mind-numbing stoop labor that most people go to college in order to avoid. Loggers shudder when they contemplate the rigors of tree planting. It is winter work, cold and wet and mud-spattered grubbing on steep mountain sides. It has always been done by those at the very lowest levels of the Northwest social order, hillbillies, drifters, derelicts, migrant farm workers and illegal aliens. And yet, a survey of the Hoedads Inc. membership in the late 1970's found that planters with post-graduate degrees were more common among them than high school drop-outs.

Forty years ago, when a Lane County, Oregon tree planting crew named itself after their distinctive planting tool, many young people believed, quite seriously, in creating a new approach to living. The "Counterculture" it was called, and though it presented itself in forms that were shocking to their parents and to the House Un-American Activities Committee it was, at heart, just an attempt to bring into the world a society that lived up to solidly American principles. Beneath the beards, beads, long hair, and odd forms of dress and speech, the hippies were merely young people who wanted to live according to the sorts of things they'd been brought up to cherish: freedom, equality, kindness, honesty—all the noble Sunday School and scouting values that, as children, they'd been taught to believe in, and which, they later discovered, were so very often either ignored or routinely violated in the conduct of our nation's governance and business practices.

When the old-time hippies of old spoke you could almost hear the initial capital letters for certain words and phrases. They called themselves Freaks. They wanted to Expand Consciousness and to Live Authentically, outside of The System, without Working For The Man. They all understood these terms, all of which, I suppose, must seem quaint and a bit puzzling to their children and grandchildren nowadays. Two of the most common approaches to supporting yourself while living an honorable Alternative Life-Style were starting your own small business and moving out of the city to a place out in the boondocks. "Right Living" and "Back To The Land" these notions, which became movements, were called.

Unfortunately, meticulously making handicrafts on backwoods communes seldom paid well-enough to feed a family. Other work was difficult to obtain because most of the bosses hated the hippies almost as much as the hippies hated having bosses. What little work could be had was mostly in doing the harsher forms of manual labor—picking fruit, bucking hay, cutting fire wood, or planting trees—work that "respectable" folks left to society's outcasts and outlaws. It was a youthful and romantic time. The Freaks were eager to cast themselves out and living outside of oppressive laws that they didn't respect was only the righteous thing to do.

Jerry Rust and John Sundquist got their first job planting trees in December of 1969. Rusthad graduated from the University of Oregon in 1965 with a degree in political science and, after serving a two-year stint in the Peace Corps, returned to Eugene and married Sidney Roscoe.

Sundquist was an Honors student at the University of Oregon who wasn't particularly interested in his course work. The two hired on with a reforestation contractor planting on Weyerhauser land up Fall Creek, earning $3.25 per hour. They worked for the contractor through the winter planting season and, having noticed the disparity between what they earned and what the contractor was receiving for their work, decided to bid their own contracts for the winter of 1970–1971.

With their friend John Corbin, they bid on a Bureau of Land Management contract of their own, sixty-three acres of steep ground up on Humbug Mountain in the Coast Range near Port Orford. It didn't go well. The land was steep, the weather bad and living in a tent through the stormy coastal Oregon winter was even less fun than you might expect. However, the contract was completed and a second planting contract in the early spring went better. By the end of the season the worker-owner trio had earned $2,700 for five months-worth of work. It wasn't a princely sum—a good deal less than what they'd have made Working for The Man on a reforestation contractor's crew—but for people who took pride in having dropped out of our money-mad society, it wasn't bad either. And what could be better Karma than planting trees?

That summer they bid new contracts for the following winter season and expanded their partnership to eight worker-owners. The three had been calling themselves The Triads, but with more partners that became obsolete. Someone suggested naming themselves after their brutally efficient planting tool, the hoedad, and the name stuck.

Each year brought more planters and more contracts. The first Hoedad crew grew to an ungainly

50 planters before dividing itself, amoeba like, into smaller, more workable groups. Shortly after the first stages of growth the question of just how big the co-op could get and still be self-manageable by fully equal and fully engaged partners came up. One hundred? What if there were say, two hundred or maybe even five hundred worker-owners deciding things? A call for workers went out, crews were hastily formed and partially trained and sent out on large contracts. By 1974, in large part through the efforts of Ed Wemple, Hoedads had incorporated as a full-fledged cooperative and boasted nine crews. The total contract earnings during those first five years came to $2,395,491 in 2009 US Dollars.

At their peak, in the late 1970's Hoedads, Inc. had about 250 members and annual earnings over $6 million (adjusted) per year. With all of the joining and quitting, some 3,000 men and women worked as Hoedad partners over the co-op's twenty-four year life-span. Of those three thousand planters many left the co-op voluntarily (some within just a day or two of hitting the slopes) but none of them were ever forced to quit.

From time to time a Hoedad or two, or three, would show up on one of the company planting crews that I used to work on. They would plant for a couple of weeks or a month or two while in-between their own co-op contracts. They were always good tree planters and good people to work with. There was an easy confidence about them, an air of amused invulnerability when dealing with the foreman. They came bearing subversive tales of fantastic earnings that were double or triple our wages. They joked a lot out on the units, teased every authority figure in sight and when they'd leave we missed them.

The Hoedads were real hippies. They weren't television and movie hippies—all flowers and headbands and incense—but actual funky, fiercely independent and often downright ornery Freaks, who were also idealistic and compassionate almost to a fault. They made many mistakes early on but learned and adjusted as they grew from a handful of unemployed friends to a large anarcho-syndicalist cooperative enterprise with an annual gross income that any hard-shell capitalist would envy. Developing ways of working together took a great deal of hard work in itself. It helped that there was money involved, since money provided a solid reason to organize. What helped the most though was compassion, the belief that everyone was truly equal and deserved to be treated as everyone else's equal, not just in theory, but in fact. In large measure the co-op ran on respect as much, or more, than it did on manual labor and on money.

Joy was always a big part of the Hoedad scene. Compassion does not merely lead to celebration, it demands it, and the Hoedads, if nothing else, knew how to celebrate. They were always ready to sit in a sweat lodge, soak in a hot-spring or skinny-dip in a creek at the drop of a pair of blue jeans. Whole hogs, kegs of beer and hailstorms of killer-bud reefers disappeared at Hoedad crew end-of-season parties and the rising sun would reveal a flattened circle of meadow grass pounded flat and shiny by barefooted dancers. The 1976 Hoedads Inc. annual general meeting took place along the South Umpqua River, upriver from Tiller at the Johnny Springs teepee burner. It was summer and it was hot and so, naturally, their business meeting took place in the river and in the buff.

Both large-scale corporate capitalism and bureaucratic socialism as practiced during the Twentieth Century lacked compassion, joy and affection—the very things that make human life worth living. It was a remarkably inhumane century filled with massive crimes against humanity committed in the names of both capitalism and socialism. East or West, left or right, the common destructive feature was the lack of what people have always treasured the most, humane and loving relations with each other and with the earth. During the final days of the Union of Soviet Socialist Republics' social and political congestive heart failure there was a too-late promise of "socialism with a human face," the lack of which had already brought the country to the brink of dissolution. In large measure, what the Hoedad experiment aimed at was developing a workable form of capitalism "with a human face."

In the process of earning their honest living they helped change their world in many ways. One simple concept that the Hoedads helped introduce to reforestation was the notion of doing high quality work, to insure a high survival rate for the seedlings they planted. The contractors who dominated the field prior to the Hoedads coming, had little or no interest in what happened to the land after they'd completed their contracts. "Jump in; blow it out; move on," was about the extent of their commitment. Planting was done on a piece-work basis in a careless rush, skipping areas wherever possible and burying trees to increase the payments. Foresters were puzzled and a bit alarmed at first when the Hoedads started turning in excess trees after finishing a unit. Honesty just wasn't part of the business before these hairy misfits arrived on the scene.

Hoedad crews had about 25% women among their worker-owners, a novelty in what had been a strictly male world of reforestation. It had frankly never occurred to any one that a woman could plant trees efficiently or ever would consider it to begin with. Yet women came to Hoedads and they planted competently—sometimes while pregnant—and were seriously committed to supporting their fellow crew members.

Hoedad crews planted trees, dug fire lines, thinned timber stands, built bridges and fences and cut forest trails in every state west of the Rockies and in Canada, Alaska and Hawaii. They also helped spawn other forest work cooperatives, lending their organizational know-how, start-up capital and even temporary Hoedad membership to new co-ops.

With economic success they became politically powerful. The Hoedads successfully fought the use of herbicides and pesticides in reforestation. When their members started getting sick from exposure to Thiram, a toxic chemical coating applied to the seed-ling trees in order to discourage browsing by wildlife, they fought to have the substance banned for use on seedling trees. As a local political force they hit their peak with the election of Hoedad President Jerry Rust as Lane County commissioner in 1976, a post he held for the next twenty years.

Eugene, Oregon's reputation as one of America's most liberal cities, the equal of Berkeley, California, Ann Arbor, Michigan and Madison, Wisconsin owes much to the Hoedads. It is virtually impossible to name a Lane County countercultural institution that doesn't, in some way, owe something to either Hoedads Inc. or to some Hoedads. The W.O.W Hall, Growers' Market, Saturday Market, and the Oregon Country Fair all ben-efited from either Hoedad money, or Hoedad expertise. Hoedad charitable giving was common, wide-spread and generous to social and political activist groups.

The heyday of the Hoedads lasted about a decade, roughly from 1974 to 1984. The early 1980's brought the "trickle-down" recession to the Northwest.

Nationwide, the number of housing starts plummeted drastically and with it the sale of lumber which, in turn, meant less logging and less timber-cutting and therefore fewer acres of land in need of reforestation. The bidding for fewer and fewer contracts became much more competitive and the contracts themselves were generally for smaller acreages. Hoedads Inc. annual contract earnings dropped to about one third of their pre-recession levels.

At the same time, many of the long-time members found themselves older and looking for work that involved less wear and tear on their ageing bodies. Much of the original spirit died out with their departure. The attitude changed from that of comrades in subversion to something more like workers with a union hiring hall. By the early 1980's the co-op was debating the hiring of employees. The co-op eventually stopped doing reforestation altogether, took up construction work and hired hourly-wage earning employees. By then, most of the "old-timer" worker-owners had moved on to other co-ops or gone back to college and on to careers in medicine, real estate, teaching and other ways of earning one's keep that didn't involve performing stoop labor on steep mountainsides. Following a large cost-overrun on a wooden bridge restoration contract, the Board of Directors of Hoedads Incorporated voted to pull the plug on the co-op in 1994.

Hoedads Inc. is gone now, but most of the Hoedads themselves remain among us. I always smile and relax upon finding that the person I'm talking to has planted trees for a living. It's as if we've known each other for a long time and have so much that we don't need to say to each other.

Sometimes I think that tree planters are the only real people.

2011

There's a stand of douglas fir trees near I-5 halfway between my place in Myrtle Creek, Oregon and my brother's house in Eugene. I was on the crew that planted them thirty-six years ago and I remember sitting up on the hillside at break time, looking down on the interstate and thinking about how I'd be able to watch them grow by looking up at them every time I drove up north.

Plans like that seldom work out but this one did. Five or six times a year I find myself leaving the Hundred Valleys of the Umpqua on my way to the Emerald Empire and halfway there I check on the trees.

For the first four years or so the seedlings were invisible from the highway. The hill looked like what it was, some slopes cleared to make pasture for a sheep ranch with nothing to see but green grass and black stumps. When the trees were finally big enough to be seen, tiny dots in rows crossing the hillside, I wasn't sure whether I was seeing them or not. I stopped by the roadside to make sure, got out of my pickup and squinted up at them while log trucks boomed by behind my back.

Over the years I've pointed out the trees to friends and family and hitch hikers who've ridden with me. In five seasons during the late 1970s I planted tens of thousands of trees in remote places, hours away from the freeway on labyrinthine logging roads that snake through the mountains. So it's been a source of pride for me to watch this one forty-acre patch of renewed forest. There aren't too many things I've done in my life whose effects I could watch over the years like that, a green and living legacy.

―――

Some time during the George H. W. Bush years I noticed an addition to the scene, one of those little crop signs, like the ones the state puts up for the benefit of people who can't tell the difference between a wheat crop and a hayfield. This one read: "TREE FARM. PLANTED 1976."

"(By Hippies and Hillbillies)," I wanted to add.

Seeing that sign was an unsettling experience. I was ambushed by an unruly mob of memories and emotions, some of them conflicting and some simply disjointed, but all of them powerful.

One source of my consternation was seeing the words "tree farm" applied to the hillside plantation. I don't believe in tree farms or the mentality which uses the term. I believe in forests—complex communities of trees, plants, critters and people, where the playful but deadly serious work of evolution goes on and on, " . . . world without end. Amen." A tree farm is not a forest but an industrial site.

Industrial silviculture is like Longfellow's little girl with her little curl; when it's good it's very, very good but when it's bad, it's horrid. When it works, as it has on that hillside, the results are stunning. Unfortunately, it has its failures too. I've worked on thousands of acres of them over the years.

Even silviculture's successes are only short-term, something the Germans, who invented the techniques and attitudes for it, have learned. Their tree farms are dying and they have no unmanaged forests left. Most German foresters have never seen a real forest—they come here to Oregon to try to figure out where they went wrong.

It's easy for people to mistake a plantation such as this one for an actual forest. After all, it has something of the appearance of a forest, being a hillside thickly covered with conifers. But the resemblance ends there. It hasn't the varied and somewhat ragged look of a forest and it doesn't provide the same things that a forest provides in the way of fish and wildlife habitat, resistance to wildfire or water storage and filtration.

At the time of our planting back in 1976 both the government and the timber industry planned to convert virtually all of the standing forests in Douglas

County, Oregon into maximum production tree farming units. The Federal government's decision to manage our forests as real forests rather than as industrial raw material extraction and production sites was vigorously opposed by the industrialists. By and large, they refuse to this day to accept the scientific research that proved that plantation-style forest management is unsustainable and harmful to the land and to the land's critters.

---

Those 20,000 little foot-tall seedlings that we planted decades ago are now about fifty feet tall and will be cut down—perhaps as soon as four or five years from now. It's an odd feeling to see them standing tall and dark up on that hillside and to know that one day, probably within the next ten years, I'll look up there and see stumps and logging slash.

There is a pride that good workers feel in the results of their work, a sense of ownership in the things they've made. It reaches poignancy sometimes when the emotional attachment without which good work is impossible is to something that belongs to others. Despite what my heart tells me, "my" trees are not really mine at all. It is, I suppose, foolish to care for decades about an industrial site where I once spent a spring-time week working, but then, caring is what people seem to do best and what all but the sociopaths among us find makes life worth living.

Of course, I knew full well when I planted those seedlings that I was helping to create a plantation. But I am sixty years old now and no longer twenty-four and in the intervening years life has become more precious to me—even the life of trees—as so many whom I knew have passed on. Silly as it may sound, the harvest of these trees will, I'm sure, set me to mourning as though for a long-time friend.

2012

# BARN RAZING

I was tearing down my barn a few weeks back when I was overwhelmed by the oddest feeling—not of nostalgia, nor of sadness—but of calm, grace and great peace. I had been working at it slowly, unhurried and inefficiently, removing each board singly and then pulling all of its nails, tossing the nails into an old coffee can and then stacking the board neatly with the rest several feet away. Having labored professionally for over forty years—often-enough under bosses who insisted on seeing "nothing but asses and elbows" at work—it felt soothing to be approaching the job as an art, to be done not merely in order to hang on to a job, but rather, appreciatively, as a useful and necessary ritual.

God knows, the about-to-tumble old wooden eyesore held enough of my memories to fill a Russian novel. I had thought of them singly and in bunches while standing out there up on ladders, with hammer and pry bar in hand. I had milked Marigold the dairy cow through four lactations in that building and bottle-fed four of her wobbly-legged calves there. Here and there, in odd corners I found remnants of my grown son's childhood toys, some broken plastic pieces of a Batmobile, and a small one-legged disabled veteran named GI Joe. "Kurt-ifacts" I called them, feeling like an archeologist might when digging into his own past.

For several years the barn had been a place of my daily work: milking the cow by hand morning and night nine months of the year, feeding her, her calves and a mother-and-son pair of cats, emptying and filling the hay loft. But that was over two decades ago and the barn had fallen into disuse and then disrepair and it was time to pull it all down, before gravity and wind knocked it down for me.

Friends and neighbors urged me to simply "torch it off" by tossing a lit road-emergency flare into the ton and a half of remaining hay up in the loft, but I insisted on dismantling it instead. I told them that I wanted to re-cycle and re-use whatever I could of the rough-cut lumber, poles, roof-tin and decades-old hay. Besides, I pointed-out, dismantling the building would make for less of a mess to clean up. They would nod and admit that doing so would be a more ecologically sound practice, even if it was obviously much more trouble than it was worth. This "keep it green" line of mine was simply a dodge though. I think, really, that it just felt disrespectful to me to set the poor thing ablaze.

I know that seems odd, to talk of respect for a mere building, but then I have worked off and on in the construction trades for many years and have come to believe that there is more to a house (or barn or shed) than just the basic elements of wood and iron and concrete and glass. As such, the buildings where we spend our days and nights working and living deserve a bit of care, not just on financial or ecological grounds but because they have been places where people have struggled with their hearts.

One day a friend and I grabbed our work gloves and chainsaws and drove up Louis Creek in his pre-WWII "deuce-and-half" Chevrolet stake-bed truck to scavenge poles from logging slash. We had no permit to gather wood from the lumber company's clear-cuts and (being members in good standing of the hickory shirt, dungarees and suspenders fraternity) we would have been deeply insulted if anyone had suggested that we needed anyone's permission in order to glean building materials or fuel from a corporation's mountainside burn-piles. We were after "worthless" poles, Douglas fir under six-inches in diameter and Incense cedar up to ten inches for the barn's frame. A gyppo sawmill operator owed me some payroll money that he was having a hard time coming up with and I traded him what was owed on my labor for enough 1x4 skip sheathing and rough-cut 1x12 planking to provide a roof deck and hayloft floor.

My barn rose up one Saturday in a half-drunken hurry, tossed up in a youthful fit of funky anti-crafts-manship. The motto for the day was, "Screw it! That's good enough," a phrase that I used much too often as we assembled the pole framing. I cut some half-assed joints with a chainsaw and we spiked the bark-clad poles together with 20 penny nails. At the end of the day I was proud of not having used a spirit level or plumb bob to check the work. Everything had been set by eyeball only and looking at it the next morning it was obvious that my eyes had not seen too well. I told myself that the finish work would conceal the odd frame lines once it was done.

Unfortunately, I never finished it either. I took it just far enough to keep the hay loft and milk shed dry and then left it standing unfinished for nearly thirty years. It really did end up, to my lasting shame, as a squalid sight and one that didn't improve any with age. In the end it seemed that, since it had been tossed together with so little regard, I was obligated to demolish it respectfully.

A neighbor of mine has spent a year and a half digging up rocks around his place and stacking them into long wide dry-stone terrace retaining walls. He's kept at it nearly every day, through summer heat and winter rains and he must have hand-picked, moved and laid a dozen tons of rock using only a shovel, a pry-bar, a pick and a wheelbarrow. I've got to admire that, though I suspect that, like many gray-haired former laborers, stoop work had for him become normal and habitually tolerable over the decades while his newly-won retirement leisure was unbearable. Over the years, several elderly blue-collar men have confessed to me that they believe continued daily physical work is the only thing keeping them alive. So perhaps my neighbor simply fears death more than grubbing-up stones.

And I, in my own way, am also given to relying on the familiar comfort of work gloves, hammer, shovel, pick, wheelbarrow and chainsaw. These simple tools I understand. I know them and trust them and believe in their efficacy. For decades I have, in a small way, shaped the world with my work. Now, nearing retirement age myself, I suspect that it was the world and its work that shaped me.

There's a white scaly patch of skin just below my knee now and another on the calf. My doctor tells me that it's a simple case of psoriasis, incurable and due to ageing. Soaking in the bathtub and eyeing it I was suddenly reminded of our river's sore-head Spring Chinook salmon. All summer long they wait for the fall rains, holding in deep pools, swimming in circles, with fuzzy white patches scattered here and there on their scaly skins. I too, it seems, await the end of my own journey.

I thought of all these things often during my weeks of mostly solitary demolition, picking through the wreckage of my mind and heart as I dismantled the barn. There was a sort of grieving to it at times. I would be reminded of those years of hardship and struggles and a great sadness would come over me. I felt so very sorry for the man whom I was back then and for the people whom we were, a family and a circle of friends trying hard to get by and mostly failing at it. And then, one day, I was pulling nails and thinking about the friends I'd known who have since died, the people who were connected in my memory to that now-sagging barn and my mind strayed back in time, to earlier generations in my family and I thought about how little I knew of their struggles. They, I realized, were forgotten and that to forget and to be forgotten is a blessing, an ultimate and assured peace. Someday, I too will take my own place in the Great Forgetting and the struggle will finally be at an end

2010

## WHY WE CELEBRATE

We made our three-hundred-dred-and-sixty-third, and final, monthly land payment, just a few weeks ago. Well, actually, I puttered about in the kitchen while my sagacious wife, Diane, wrote out the check and addressed the envelope. It was a good feeling to know that after thirty years and three months—a full third of a century—the old shack by the railroad tracks and two acres of serpentine soil was wholly ours. Seeing her sitting there at the dining-room table two thoughts occurred to me nearly at the same moment.

*"Sine qua non,"* I thought first, "Without whom nothing." We never could have made it through the past thirty-eight years without each other. And then, sudden-like, a shotgun-toting TV hillbilly appeared in my imagination. I looked at my dearest, frowned at her, and in my best pseudo-Ozarkian accent deadpanned an old movie cliché, "This here's Heilman land mister." It worked; she smiled.

I'd had my doubts about the place when I first saw it. It seemed too exposed, too small, too thick with yellow star thistle and poison oak. But then, at one time the landscape of Oregon seemed downright creepy to me. In the spring of 1970, about a week after the Kent State University shootings, I passed through the state after living in northern New Mexico for several months. Being accustomed to walking through the austere open high-desert country, this tall and green land felt threatening, closed-in, a dark and dampish place where things rot and decay. The desert's scattered and small plants had seemed beautiful, little individual protests against an arid death, heroic somehow in their struggle to survive. Here in western Oregon I saw a tangled forest feast of living things busily devouring the dead and each other. It just didn't seem right—the vegetation was predatory.

I couldn't escape the feeling that this overly verdant land might eat me too and hide my skinny young corpse beneath a dimly lit blanket of green oxalis and red duff. It seemed a frightening thought back then, even though, at age eighteen, I felt sure that I would live forever. Now, with the passage of time, the notion of lying beneath blankets of hardwood leaves and evergreen needles forever comforts me.

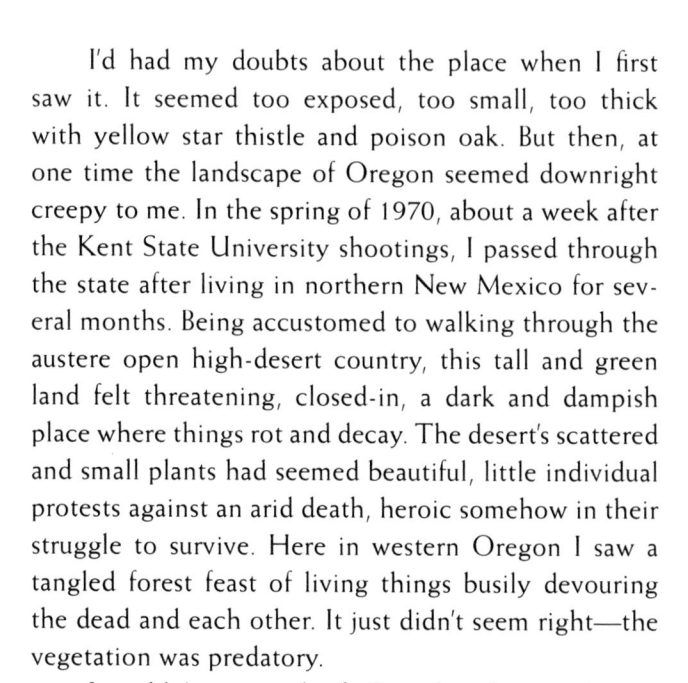

Diane and I went to a housewarming the other day. Shannon's place up at the north end of the county has been in her family since the 1840s and the event celebrated the renovation of the newer of the property's two houses, this one a mere 120 or so years old. She and Daniel had done a wonderful job but it was a hot day and a crowded house and I took my sweetheart for a short walk down to "The Old Place" built in the mid-1850s, back in the pre-Civil War Oregon Territory times.

There is a certain sort of beauty to old homesteads, one that accrues over the long years of human habitation. After the early mistakes have been corrected and generations have added to their comfort and shaped the landscape to fit their ways, the place itself matures, though slowly, much more slowly than the shorter lives of those who shape it. With maturity comes real character, for the older it gets and the longer that people live there, the more it becomes distinctly itself.

By contrast our own wholly owned piece of Oregon has barely begun to adjust to the land and to time. We live in a 1968 Nashua trailer, already dangerously obsolete by any building code standards and probably a good ten years past what anyone expected of its planned lifespan. It is the modern equivalent of a sod homestead shack of the sort that used to house my grandparents, a temporary solution to the need for a roof over our heads, run-down now and not much longer for this world. The trailer was moved here to this hillside in 1972, onto a thin marginal strip of ground wedged between the county road and the railroad tracks. I hope that the next house will fit our needs and the needs of the land better than the old one. After three decades here on the place I know that I understand this place more fully and I hope that time has given me the insightful wisdom to do better.

The phone rang the other night and the caller turned out to be a pollster, mispronouncing my surname and eager for answers to the questions of the day.

"Do you think that Oregon is heading in the right direction?" he asked first without so much as a neighborly "How do you do?"

I couldn't come up with an honest answer because a sudden swarm of impertinent questions hovered about me. "Heading" somewhere? The whole state of Oregon is going someplace? Why wasn't I told about it? Come to think of it, I had heard that it was heading due east at about 500 miles-per-hour along with everything else at this latitude. Then there's the other direction, westward with the drifting continental plates. And then, given the erosion rate, it is true that this entire county is heading downriver to Reedsport. And none of these physical movements are of much concern to me.

Of course, I knew that he was talking figuratively, as if the passage of time in Oregon and the occurrence of social and political changes here were aspects of an actual journey, which it clearly is not. But if it were, well, what then? Some things Oregonian seem to me to be going quite well, others not so well. It was all much too complicated. What could I honestly say? I had reached a state of mental overload in a matter of seconds.

"Dude," I pointed out, "you went and skipped a step—you never asked me whether I want to answer your questions or not."

"Oh . . . Well, then, would you be willing to take part in the survey?"

This second question was much easier to answer honestly, "No, not really."

I'm not sure what people mean when they say "Oregon." It seems simple enough—a geographical designation for a part of the earth and for the 150-year-old political entity spoken of in quaintly ornate rhetoric as "The Great State of Oregon"—but I can't really bring myself to care much about either of those. I suppose that, like myself, when most Oregonians say "Oregon" what they mostly mean is "my home."

People cannot love abstractions (though they may be fond of speaking in them). Even what isn't abstract but is merely remote cannot be as effectively loved as what is near. There is a sort of geographical hierarchy of affection which is natural to all humans. I can not, for example, honestly say that I love Delaware, a place I've never been to and which I have never desired to visit—yet one to which I have very real historic, economic, political, cultural and other societal ties. Love—the act itself rather than the associated emotion—requires personal engagement. At best I can include the "Great State of Delaware" within a generalized patriotism that embraces the nation itself.

Of course, whenever I return across the state line back into my Oregon I feel a sense of relief and a contentment that I can only feel here. But then, I feel it again, only stronger, when I enter Douglas County and see any of my beloved Umpqua River valleys and strongest yet when I turn down my own driveway. Within me my allegiances lie in layers like a set of those Russian nesting-dolls: my little bit of land enclosed by Orchard Valley which contains it; the valley contained within the county; the county within the state; the state within the nation; and on out toward the world itself.

Oregon has reached the venerable age of 150 years and we have lived on our place for 30 years. What is it that we celebrate when we commemorate an anniversary if not some large or small triumph over entropy, chaos, and death? The mere passage of time is not something worthy of celebration. In itself it is only inevitable and unremarkable. We are happy because we continue to exist in spite of all the things that could have destroyed what we love—each other and our place here. So many things can pull apart a state, a society, a marriage, and a family that we feel lucky to have survived.

Death is our constant companion, sitting patiently by us all of our lives, teaching us to appreciate life and each other, bringing us to understand the roots of compassion. In the end, each of us must die. Someday, all that we have worked for will be lost and forgotten. Someday this Oregon we've loved so carefully for a century and a half now will no longer exist. This is what makes it worthy of our affection, that it, like us, is transient, and our triumph only temporary.

Appreciating our luck is a part of this urge to celebrate, but only a part. We also celebrate the necessary good work without which there would be no surviving. Skillful work, attentive work, informed work, lovingly done work: these are essential aspects of both a well-lived life and of a truly viable culture. It is not through the great enterprises nor as a mass that we do good in the world but by individual small, immediate, and informed acts. At heart, it is caring that keeps an individual sane and that keeps a family intact and a society from collapse. What is a well-run state if not a place where people can and do care about each other and about their land?

People, it seems, are overwhelmingly decent—otherwise there would be no continuation of life. It is the fact of continuing survival that gives us hope. For many years I have been fond of telling people that I'm always optimistic in the long run and pessimistic in the short run. Everything always seems to be heading straight off to Hell in a handcart when looked at on a day-to-day basis.

Democracy, like marriage, is often a messy, uncomfortable, and doubtful way to manage things. But, looked at over the long haul, it somehow works out to be a very good thing for most people. We learn, over time, to be more decent in our treatment of each other. Oregon's 1859 constitution provided for whipping black Americans for the crime of residing within the state for longer than two years and also denied all women the right to vote. We learn how to do things better, and that is cultural change, from which political change follows.

Our road is old—older than both of us put together—though not as old as the hillside it crosses. We walk, two aging lovers who are friends, two Golden Retrievers who are sisters and, trailing behind, a half-Siamese kitten who adores the dogs. We walk through tunnels of bright early fall leaves, yellow big-leaf maple and now-golden oaks and below them, the bright red of poison oak back-lit by the glow of late afternoon sunlight.

As we walk it occurs to me that I can no longer walk this road without being in sight of some place or another to which we've attached stories: here we found a kitten, there a naked midnight hitchhiker stood, and beyond that the slope where our errant cow was wandering, the places where various teenagers have driven over the bank, old landslides, and the fallen trunks of once-tall trees. Overlaid on the road of today is another, a road of memories, of old desires, of triumphs and tragedies and both roads stretch on. We walk hand-in-hand not just through space but through time as well and, once in a great while, we pause to consider just how unlikely we are.

2009